Learn

Eureka Math®
Grade 5
Modules 3 & 4

Published by Great Minds®.

Copyright © 2018 Great Minds®.

Printed in the U.S.A.
This book may be purchased from the publisher at eureka-math.org.
10 9 8 7 6 5 4 3 2

ISBN 978-1-64054-072-9

G5-M3-M4-L-05.2018

Learn ◆ Practice ◆ Succeed

Eureka Math® student materials for *A Story of Units*® (K–5) are available in the *Learn, Practice, Succeed* trio. This series supports differentiation and remediation while keeping student materials organized and accessible. Educators will find that the *Learn, Practice,* and *Succeed* series also offers coherent—and therefore, more effective—resources for Response to Intervention (RTI), extra practice, and summer learning.

Learn

Eureka Math Learn serves as a student's in-class companion where they show their thinking, share what they know, and watch their knowledge build every day. *Learn* assembles the daily classwork—Application Problems, Exit Tickets, Problem Sets, templates—in an easily stored and navigated volume.

Practice

Each *Eureka Math* lesson begins with a series of energetic, joyous fluency activities, including those found in *Eureka Math Practice.* Students who are fluent in their math facts can master more material more deeply. With *Practice,* students build competence in newly acquired skills and reinforce previous learning in preparation for the next lesson.

Together, *Learn* and *Practice* provide all the print materials students will use for their core math instruction.

Succeed

Eureka Math Succeed enables students to work individually toward mastery. These additional problem sets align lesson by lesson with classroom instruction, making them ideal for use as homework or extra practice. Each problem set is accompanied by a Homework Helper, a set of worked examples that illustrate how to solve similar problems.

Teachers and tutors can use *Succeed* books from prior grade levels as curriculum-consistent tools for filling gaps in foundational knowledge. Students will thrive and progress more quickly as familiar models facilitate connections to their current grade-level content.

Students, families, and educators:

Thank you for being part of the *Eureka Math*® community, where we celebrate the joy, wonder, and thrill of mathematics.

In the *Eureka Math* classroom, new learning is activated through rich experiences and dialogue. The *Learn* book puts in each student's hands the prompts and problem sequences they need to express and consolidate their learning in class.

What is in the Learn book?

Application Problems: Problem solving in a real-world context is a daily part of *Eureka Math*. Students build confidence and perseverance as they apply their knowledge in new and varied situations. The curriculum encourages students to use the RDW process—Read the problem, Draw to make sense of the problem, and Write an equation and a solution. Teachers facilitate as students share their work and explain their solution strategies to one another.

Problem Sets: A carefully sequenced Problem Set provides an in-class opportunity for independent work, with multiple entry points for differentiation. Teachers can use the Preparation and Customization process to select "Must Do" problems for each student. Some students will complete more problems than others; what is important is that all students have a 10-minute period to immediately exercise what they've learned, with light support from their teacher.

Students bring the Problem Set with them to the culminating point of each lesson: the Student Debrief. Here, students reflect with their peers and their teacher, articulating and consolidating what they wondered, noticed, and learned that day.

Exit Tickets: Students show their teacher what they know through their work on the daily Exit Ticket. This check for understanding provides the teacher with valuable real-time evidence of the efficacy of that day's instruction, giving critical insight into where to focus next.

Templates: From time to time, the Application Problem, Problem Set, or other classroom activity requires that students have their own copy of a picture, reusable model, or data set. Each of these templates is provided with the first lesson that requires it.

Where can I learn more about Eureka Math resources?

The Great Minds® team is committed to supporting students, families, and educators with an ever-growing library of resources, available at eureka-math.org. The website also offers inspiring stories of success in the *Eureka Math* community. Share your insights and accomplishments with fellow users by becoming a *Eureka Math* Champion.

Best wishes for a year filled with aha moments!

Jill Diniz

Jill Diniz
Director of Mathematics
Great Minds

The Read–Draw–Write Process

The *Eureka Math* curriculum supports students as they problem-solve by using a simple, repeatable process introduced by the teacher. The Read–Draw–Write (RDW) process calls for students to

1. Read the problem.
2. Draw and label.
3. Write an equation.
4. Write a word sentence (statement).

Educators are encouraged to scaffold the process by interjecting questions such as

- What do you see?
- Can you draw something?
- What conclusions can you make from your drawing?

The more students participate in reasoning through problems with this systematic, open approach, the more they internalize the thought process and apply it instinctively for years to come.

Contents

Module 3: Addition and Subtraction of Fractions

Module 4: Multiplication and Division of Fractions and Decimal Fractions

Topic G: Division of Fractions and Decimal Fractions

Topic H: Interpretation of Numerical Expressions

Grade 5
Module 3

15 kilograms of rice are separated equally into 4 containers. How many kilograms of rice are in each container? Express your answer as a decimal and as a fraction.

_____

_____

_____

_____

Read   **Draw**   **Write**

Name _____   Date _____

1. Use the folded paper strip to mark points 0 and 1 above the number line and $\frac{0}{2}$, $\frac{1}{2}$, and $\frac{2}{2}$ below it.

<-->

Draw one vertical line down the middle of each rectangle, creating two parts. Shade the left half of each. Partition with horizontal lines to show the equivalent fractions $\frac{2}{4}$, $\frac{3}{6}$, $\frac{4}{8}$, and $\frac{5}{10}$. Use multiplication to show the change in the units.

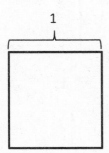 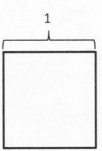 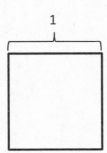 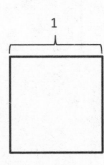

$$\frac{1}{2} = \frac{1 \times 2}{2 \times 2} = \frac{2}{4}$$

2. Use the folded paper strip to mark points 0 and 1 above the number line and $\frac{0}{3}$, $\frac{1}{3}$, $\frac{2}{3}$, and $\frac{3}{3}$ below it. Follow the same pattern as Problem 1 but with thirds.

<-->

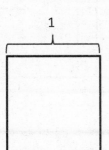 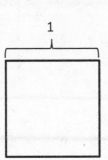 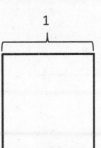 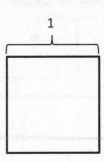

3. Continue the pattern with 3 fourths.

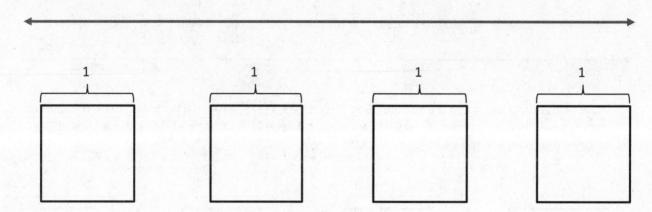

4. Continue the process, and model 2 equivalent fractions for 6 fifths.

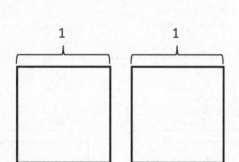   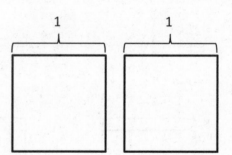

Lesson 1:   Make equivalent fractions with the number line, the area model, and numbers.

EUREKA
MATH

Name _____ Date _____

Estimate to mark points 0 and 1 above the number line, and $\frac{0}{6}$, $\frac{1}{6}$, $\frac{2}{6}$, $\frac{3}{6}$, $\frac{4}{6}$, $\frac{5}{6}$, and $\frac{6}{6}$ below it. Use the squares below to represent fractions equivalent to 1 sixth using both arrays and equations.

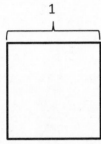  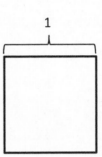

$$\frac{1}{6} = \frac{1 \times 2}{6 \times 2} = \frac{2}{12}$$

EUREKA
MATH

Mr. Hopkins has a 1-meter wire he is using to make clocks. Each fourth meter is marked off and divided into 5 smaller equal lengths. If Mr. Hopkins bends the wire at $\frac{3}{4}$ meter, what fraction of the smaller marks is that?

_____

_____

_____

Read   **Draw**   **Write**

EUREKA MATH

Lesson 2:   Make equivalent fractions with sums of fractions with like denominators.

© 2018 Great Minds®. eureka-math.org

9

Name _____   Date _____

1. Show each expression on a number line. Solve.

    a. $\frac{2}{5} + \frac{1}{5}$ b. $\frac{1}{3} + \frac{1}{3} + \frac{1}{3}$

    c. $\frac{3}{10} + \frac{3}{10} + \frac{3}{10}$ d. $2 \times \frac{3}{4} + \frac{1}{4}$

2. Express each fraction as the sum of two or three equal fractional parts. Rewrite each as a multiplication equation. Show Part (a) on a number line.

    a. $\frac{6}{7}$ b. $\frac{9}{2}$

    c. $\frac{12}{10}$ d. $\frac{27}{5}$

3. Express each of the following as the sum of a whole number and a fraction. Show Parts (c) and (d) on number lines.

a. $\frac{9}{7}$

b. $\frac{9}{2}$

c. $\frac{22}{7}$

d. $\frac{24}{9}$

4. Marisela cut four equivalent lengths of ribbon. Each was 5 eighths of a yard long. How many yards of ribbon did she cut? Express your answer as the sum of a whole number and the remaining fractional units. Draw a number line to represent the problem.

Lesson 2: Make equivalent fractions with sums of fractions with like denominators.

EUREKA MATH

Name _____   Date _____

1. Show each expression on a number line. Solve.

   a. $\frac{5}{5} + \frac{2}{5}$

   b. $\frac{6}{3} + \frac{2}{3}$

2. Express each fraction as the sum of two or three equal fractional parts. Rewrite each as a multiplication equation. Show Part (b) on a number line.

   a. $\frac{6}{9}$

   b. $\frac{15}{4}$

One ninth of the students in Mr. Beck's class list red as their favorite color. Twice as many students call blue their favorite, and three times as many students prefer pink. The rest name green as their favorite color. What fraction of the students say green or pink is their favorite color?

Extension: If 6 students call blue their favorite color, how many students are in Mr. Beck's class?`

Read   **Draw**   **Write**

Name _____ Date _____

1. Draw a rectangular fraction model to find the sum. Simplify your answer, if possible.

    a. $\frac{1}{2} + \frac{1}{3} =$

    b. $\frac{1}{3} + \frac{1}{5} =$

    c. $\frac{1}{4} + \frac{1}{3} =$

    d. $\frac{1}{3} + \frac{1}{7} =$

EUREKA MATH

Lesson 3:   Add fractions with unlike units using the strategy of creating
              equivalent fractions.

© 2018 Great Minds®. eureka-math.org

17

e. $\frac{3}{4} + \frac{1}{5} =$

f. $\frac{2}{3} + \frac{2}{7} =$

Solve the following problems. Draw a picture and write the number sentence that proves the answer.
Simplify your answer, if possible.

2. Jamal used $\frac{1}{3}$ yard of ribbon to tie a package and $\frac{1}{6}$ yard of ribbon to tie a bow. How many yards of ribbon did Jamal use?

Lesson 3:   Add fractions with unlike units using the strategy of creating
                              equivalent fractions.

EUREKA
MATH®

3. Over the weekend, Nolan drank $\frac{1}{6}$ quart of orange juice, and Andrea drank $\frac{3}{4}$ quart of orange juice. How many quarts did they drink together?

4. Nadia spent $\frac{1}{4}$ of her money on a shirt and $\frac{2}{5}$ of her money on new shoes. What fraction of Nadia's money has been spent? What fraction of her money is left?

Name _____   Date _____

Solve by drawing the rectangular fraction model.

1. $\frac{1}{2} + \frac{1}{5} =$

2. In one hour, Ed used $\frac{2}{5}$ of the time to complete his homework and $\frac{1}{4}$ of the time to check his email. How much time did he spend completing homework and checking email? Write your answer as a fraction. (Extension: Write the answer in minutes.)

Leslie has 1 liter of milk in her refrigerator to drink today. She drank $\frac{1}{2}$ liter of milk for breakfast and $\frac{2}{5}$ liter of milk for dinner. How much of a liter did Leslie drink during breakfast and dinner?

Extension: How much of a liter of milk does Leslie have left to drink with her dessert? Give your answer as a fraction of liters and as a decimal.

_____

_____

_____

_____

Read   **Draw**   **Write**

Name _____   Date _____

1. For the following problems, draw a picture using the rectangular fraction model and write the answer. When possible, write your answer as a mixed number.

   a. $\frac{2}{3} + \frac{1}{2} =$ b. $\frac{3}{4} + \frac{2}{3} =$

   c. $\frac{1}{2} + \frac{3}{5} =$ d. $\frac{5}{7} + \frac{1}{2} =$

e. $\frac{3}{4} + \frac{5}{6} =$

f. $\frac{2}{3} + \frac{3}{7} =$

Solve the following problems. Draw a picture, and write the number sentence that proves the answer. Simplify your answer, if possible.

2. Penny used $\frac{2}{5}$ lb of flour to bake a vanilla cake. She used another $\frac{3}{4}$ lb of flour to bake a chocolate cake. How much flour did she use altogether?

Lesson 4:   Add fractions with sums between 1 and 2.

EUREKA
MATH®

3. Carlos wants to practice piano 2 hours each day. He practices piano for $\frac{3}{4}$ hour before school and $\frac{7}{10}$ hour when he gets home. How many hours has Carlos practiced piano? How much longer does he need to practice before going to bed in order to meet his goal?

Name _____ Date _____

1. Draw a model to help solve $\frac{5}{6} + \frac{1}{4}$. Write your answer as a mixed number.

2. Patrick drank $\frac{3}{4}$ liter of water Monday before jogging. He drank $\frac{4}{5}$ liter of water after his jog. How much water did Patrick drink altogether? Write your answer as a mixed number.

A farmer uses $\frac{3}{4}$ of his field to plant corn, $\frac{1}{6}$ of his field to plant beans, and the rest to plant wheat. What fraction of his field is used for wheat?

_____

_____

_____

_____

Read   **Draw**   **Write**

Lesson 5:   Subtract fractions with unlike units using the strategy of creating
equivalent fractions.

© 2018 Great Minds®. eureka-math.org

31

Name _____   Date _____

1. For the following problems, draw a picture using the rectangular fraction model and write the answer. Simplify your answer, if possible.

   a. $\dfrac{1}{3} - \dfrac{1}{4} =$

   b. $\dfrac{2}{3} - \dfrac{1}{2} =$

   c. $\dfrac{5}{6} - \dfrac{1}{4} =$

   d. $\dfrac{2}{3} - \dfrac{1}{7} =$

EUREKA MATH

Lesson 5:   Subtract fractions with unlike units using the strategy of creating equivalent fractions.

© 2018 Great Minds®. eureka-math.org

33

e. $\frac{3}{4} - \frac{3}{8} =$

f. $\frac{3}{4} - \frac{2}{7} =$

2. Mr. Penman had $\frac{2}{3}$ liter of salt water. He used $\frac{1}{5}$ of a liter for an experiment. How much salt water does Mr. Penman have left?

Lesson 5: Subtract fractions with unlike units using the strategy of creating equivalent fractions.

© 2018 Great Minds®. eureka-math.org

EUREKA MATH

3. Sandra says that $\frac{4}{7} - \frac{1}{3} = \frac{3}{4}$ because all you have to do is subtract the numerators and subtract the denominators. Convince Sandra that she is wrong. You may draw a rectangular fraction model to support your thinking.

Lesson 5:   Subtract fractions with unlike units using the strategy of creating
equivalent fractions.

© 2018 Great Minds®. eureka-math.org

35

Name _____   Date _____

For the following problems, draw a picture using the rectangular fraction model and write the answer. Simplify your answer, if possible.

a. $\frac{1}{2} - \frac{1}{7} =$

b. $\frac{3}{5} - \frac{1}{2} =$

EUREKA MATH

Lesson 5:   Subtract fractions with unlike units using the strategy of creating equivalent fractions.

© 2018 Great Minds®. eureka-math.org

37

The Napoli family combined two bags of dry cat food in a plastic container. One bag had $\frac{5}{6}$ kg of cat food. The other bag had $\frac{3}{4}$ kg. What was the total weight of the container after the bags were combined?

_____

_____

_____

_____

Read   **Draw**   **Write**

Lesson 6:   Subtract fractions from numbers between 1 and 2.   39

© 2018 Great Minds®. eureka-math.org

Name _____ Date _____

1. For the following problems, draw a picture using the rectangular fraction model and write the answer. Simplify your answer, if possible.

   a. $1\frac{1}{4} - \frac{1}{3} =$

   b. $1\frac{1}{5} - \frac{1}{3} =$

   c. $1\frac{3}{8} - \frac{1}{2} =$

   d. $1\frac{2}{5} - \frac{1}{2} =$

EUREKA MATH

Lesson 6: Subtract fractions from numbers between 1 and 2.

41

© 2018 Great Minds®. eureka-math.org

e. $1\frac{2}{7} - \frac{1}{3} =$

f. $1\frac{2}{3} - \frac{3}{5} =$

2.   Jean-Luc jogged around the lake in $1\frac{1}{4}$ hour. William jogged the same distance in $\frac{5}{6}$ hour. How much longer did Jean-Luc take than William in hours?

42   **Lesson 6:**   Subtract fractions from numbers between 1 and 2.

© 2018 Great Minds®. eureka-math.org

EUREKA MATH

3. Is it true that $1\frac{2}{5} - \frac{3}{4} = \frac{1}{4} + \frac{2}{5}$? Prove your answer.

Name _____ Date _____

For the following problems, draw a picture using the rectangular fraction model and write the answer. Simplify your answer, if possible.

a. $1\frac{1}{5} - \frac{1}{2} =$

b. $1\frac{1}{3} - \frac{5}{6} =$

Name _____ Date _____

Solve the word problems using the RDW strategy. Show all of your work.

1. George weeded $\frac{1}{5}$ of the garden, and Summer weeded some, too. When they were finished, $\frac{2}{3}$ of the garden still needed to be weeded. What fraction of the garden did Summer weed?

2. Jing spent $\frac{1}{3}$ of her money on a pack of pens, $\frac{1}{2}$ of her money on a pack of markers, and $\frac{1}{8}$ of her money on a pack of pencils. What fraction of her money is left?

3. Shelby bought a 2-ounce tube of blue paint. She used $\frac{2}{3}$ ounce to paint the water, $\frac{3}{5}$ ounce to paint the sky, and some to paint a flag. After that, she has $\frac{2}{15}$ ounce left. How much paint did Shelby use to paint her flag?

4. Jim sold $\frac{3}{4}$ gallon of lemonade. Dwight sold some lemonade, too. Together, they sold $1\frac{5}{12}$ gallons. Who sold more lemonade, Jim or Dwight? How much more?

Lesson 7:   Solve two-step word problems.

EUREKA MATH

5. Leonard spent $\frac{1}{4}$ of his money on a sandwich. He spent 2 times as much on a gift for his brother as on some comic books. He had $\frac{3}{8}$ of his money left. What fraction of his money did he spend on the comic books?

Name _____ Date _____

Solve the word problem using the RDW strategy. Show all of your work.

Mr. Pham mowed $\frac{2}{7}$ of his lawn. His son mowed $\frac{1}{4}$ of it. Who mowed the most? How much of the lawn still needs to be mowed?

Jane found money in her pocket. She went to a convenience store and spent $\frac{1}{4}$ of her money on chocolate milk, $\frac{3}{5}$ of her money on a magazine, and the rest of her money on candy. What fraction of her money did she spend on candy?

_____

_____

_____

_____

Read   **Draw**   **Write**

Lesson 8:   Add fractions to and subtract fractions from whole numbers using
equivalence and the number line as strategies.

53

© 2018 Great Minds®. eureka-math.org

Name _____   Date _____

1. Add or subtract.

   a. $2 + 1\frac{1}{5} =$

   b. $2 - 1\frac{3}{8} =$

   c. $5\frac{2}{5} + 2\frac{3}{5} =$

   d. $4 - 2\frac{2}{7} =$

   e. $9\frac{3}{4} + 8 =$

   f. $17 - 15\frac{2}{3} =$

   g. $15 + 17\frac{2}{3} =$

   h. $100 - 20\frac{7}{8} =$

EUREKA
MATH®

Lesson 8:   Add fractions to and subtract fractions from whole numbers using
equivalence and the number line as strategies.

55

© 2018 Great Minds®. eureka-math.org

2. Calvin had 30 minutes in time-out. For the first $23\frac{1}{3}$ minutes, Calvin counted spots on the ceiling. For the rest of the time, he made faces at his stuffed tiger. How long did Calvin spend making faces at his tiger?

3. Linda planned to spend 9 hours practicing piano this week. By Tuesday, she had spent $2\frac{1}{2}$ hours practicing. How much longer does she need to practice to reach her goal?

EUREKA MATH

4. Gary says that $3 - 1\frac{1}{3}$ will be more than 2, since $3 - 1$ is 2. Draw a picture to prove that Gary is wrong.

Lesson 8:   Add fractions to and subtract fractions from whole numbers using
                equivalence and the number line as strategies.

© 2018 Great Minds®. eureka-math.org

Name _____   Date _____

Add or subtract.

a. $5 + 1\frac{7}{8} =$

b. $3 - 1\frac{3}{4} =$

c. $7\frac{3}{8} + 4 =$

d. $4 - 2\frac{3}{7} =$

EUREKA MATH®   **Lesson 8:**   Add fractions to and subtract fractions from whole numbers using   59
                              equivalence and the number line as strategies.

© 2018 Great Minds®. eureka-math.org

empty number line

Lesson 8:   Add fractions to and subtract fractions from whole numbers using
equivalence and the number line as strategies.

61

Hannah and her friend are training to run in a 2-mile race. On Monday, Hannah ran $\frac{1}{2}$ mile. On Tuesday, she ran $\frac{1}{5}$ mile farther than she ran on Monday.

a. How far did Hannah run on Tuesday?

b. If her friend ran $\frac{3}{4}$ mile on Tuesday, how many miles did the girls run in all on Tuesday?

Read   **Draw**   **Write**

Name _____   Date _____

1. First make like units, and then add.

    a. $\frac{3}{4} + \frac{1}{7} =$ b. $\frac{1}{4} + \frac{9}{8} =$

    c. $\frac{3}{8} + \frac{3}{7} =$ d. $\frac{4}{9} + \frac{4}{7} =$

    e. $\frac{1}{5} + \frac{2}{3} =$ f. $\frac{3}{4} + \frac{5}{6} =$

g. $\frac{2}{3} + \frac{1}{11} =$

h. $\frac{3}{4} + 1\frac{1}{10} =$

2. Whitney says that to add fractions with different denominators, you always have to multiply the denominators to find the common unit; for example:

$$\frac{1}{4} + \frac{1}{6} = \frac{6}{24} + \frac{4}{24}$$

Show Whitney how she could have chosen a denominator smaller than 24, and solve the problem.

Lesson 9: Add fractions making like units numerically.

EUREKA MATH

3. Jackie brought $\frac{3}{4}$ of a gallon of iced tea to the party. Bill brought $\frac{7}{8}$ of a gallon of iced tea to the same party. How much iced tea did Jackie and Bill bring to the party?

4. Madame Curie made some radium in her lab. She used $\frac{2}{5}$ kg of the radium in an experiment and had $1\frac{1}{4}$ kg left. How much radium did she have at first? (Extension: If she performed the experiment twice, how much radium would she have left?)

Name _____   Date _____

Make like units, and then add.

a. $\frac{1}{6} + \frac{3}{4} =$

b. $1\frac{1}{2} + \frac{2}{5} =$

To make punch for the class party, Mrs. Lui mixed $1\frac{1}{3}$ cups orange juice, $\frac{3}{4}$ cup apple juice, $\frac{2}{3}$ cup cranberry juice, and $\frac{3}{4}$ cup lemon-lime soda. Mixed together, how many cups of punch does the recipe make?

Extension: Each serving is 1 cup. How many batches of this recipe does Mrs. Lui need to serve her 20 students?

_____

_____

_____

_____

Read   **Draw**   **Write**

Name _____   Date _____

1. Add.

   a. $2\frac{1}{4} + 1\frac{1}{5} =$

   b. $2\frac{3}{4} + 1\frac{2}{5} =$

   c. $1\frac{1}{5} + 2\frac{1}{3} =$

   d. $4\frac{2}{3} + 1\frac{2}{5} =$

   e. $3\frac{1}{3} + 4\frac{5}{7} =$

   f. $2\frac{6}{7} + 5\frac{2}{3} =$

g. $15\frac{1}{5} + 3\frac{5}{8} =$

h. $15\frac{5}{8} + 5\frac{2}{5} =$

2. Erin jogged $2\frac{1}{4}$ miles on Monday. Wednesday, she jogged $3\frac{1}{3}$ miles, and on Friday, she jogged $2\frac{2}{3}$ miles. How far did Erin jog altogether?

Lesson 10:   Add fractions with sums greater than 2.

EUREKA
MATH®

3. Darren bought some paint. He used $2\frac{1}{4}$ gallons painting his living room. Aftewr that, he had $3\frac{5}{6}$ gallons left. How much paint did he buy?

4. Clayton says that $2\frac{1}{2} + 3\frac{3}{5}$ will be more than 5 but less than 6 since 2 + 3 is 5. Is Clayton's reasoning correct? Prove him right or wrong.

Name _____   Date _____

Add.

1. $3\frac{1}{2} + 1\frac{1}{3} =$

2. $4\frac{5}{7} + 3\frac{3}{4} =$

Meredith went to the movies. She spent $\frac{2}{5}$ of her money on a ticket and $\frac{3}{7}$ of her money on popcorn. How much of her money did she spend?

Extension: How much of her money is left?

_____

_____

_____

_____

Read   **Draw**   **Write**

Lesson 11:   Subtract fractions making like units numerically.

Name _____ Date _____

1. Generate equivalent fractions to get like units. Then, subtract.

   a. $\frac{1}{2} - \frac{1}{3} =$

   b. $\frac{7}{10} - \frac{1}{3} =$

   c. $\frac{7}{8} - \frac{3}{4} =$

   d. $1\frac{2}{5} - \frac{3}{8} =$

   e. $1\frac{3}{10} - \frac{1}{6} =$

   f. $2\frac{1}{3} - 1\frac{1}{5} =$

   g. $5\frac{6}{7} - 2\frac{2}{3} =$

   h. Draw a number line to show that your answer to (g) is reasonable.

2. George says that, to subtract fractions with different denominators, you always have to multiply the denominators to find the common unit; for example:

$$\frac{3}{8} - \frac{1}{6} = \frac{18}{48} - \frac{8}{48}$$

Show George how he could have chosen a denominator smaller than 48, and solve the problem.

3. Meiling has $1\frac{1}{4}$ liter of orange juice. She drinks $\frac{1}{3}$ liter. How much orange juice does she have left?

(Extension: If her brother then drinks twice as much as Meiling, how much is left?)

4. Harlan used $3\frac{1}{2}$ kg of sand to make a large hourglass. To make a smaller hourglass, he only used $1\frac{3}{7}$ kg of sand. How much more sand did it take to make the large hourglass than the smaller one?

EUREKA
MATH®

Name _____   Date _____

Generate equivalent fractions to get like units. Then, subtract.

a. $\frac{3}{4} - \frac{3}{10} =$

b. $3\frac{1}{2} - 1\frac{1}{3} =$

Problem 1

Max's reading assignment was to read $15\frac{1}{2}$ pages. After reading $4\frac{1}{3}$ pages, he took a break. How many more pages does he need to read to finish his assignment?

Problem 2

Sam and Nathan are training for a race. Monday, Sam ran $2\frac{3}{4}$ miles, and Nathan ran $2\frac{1}{3}$ miles. How much farther did Sam run than Nathan?

Read   **Draw**   **Write**

Name _____   Date _____

1. Subtract.

   a. $3\frac{1}{5} - 2\frac{1}{4} =$

   b. $4\frac{2}{5} - 3\frac{3}{4} =$

   c. $7\frac{1}{5} - 4\frac{1}{3} =$

   d. $7\frac{2}{5} - 5\frac{2}{3} =$

   e. $4\frac{2}{7} - 3\frac{1}{3} =$

   f. $9\frac{2}{3} - 2\frac{6}{7} =$

g. $17\frac{2}{3} - 5\frac{5}{6} =$

h. $18\frac{1}{3} - 3\frac{3}{8} =$

2. Toby wrote the following:

$$7\frac{1}{4} - 3\frac{3}{4} = 4\frac{2}{4} = 4\frac{1}{2}.$$

Is Toby's calculation correct? Draw a number line to support your answer.

Lesson 12:   Subtract fractions greater than or equal to 1.

EUREKA
MATH

3. Mr. Neville Iceguy mixed up $12\frac{3}{5}$ gallons of chili for a party. If $7\frac{3}{4}$ gallons of chili was mild, and the rest was extra spicy, how much extra spicy chili did Mr. Iceguy make?

4. Jazmyne decided to spend $6\frac{1}{2}$ hours studying over the weekend. She spent $1\frac{1}{4}$ hours studying on Friday evening and $2\frac{2}{3}$ hours on Saturday. How much longer does she need to spend studying on Sunday in order to reach her goal?

Name _____   Date _____

Subtract.

1. $5\frac{1}{2} - 1\frac{1}{3} =$

2. $8\frac{3}{4} - 5\frac{5}{6} =$

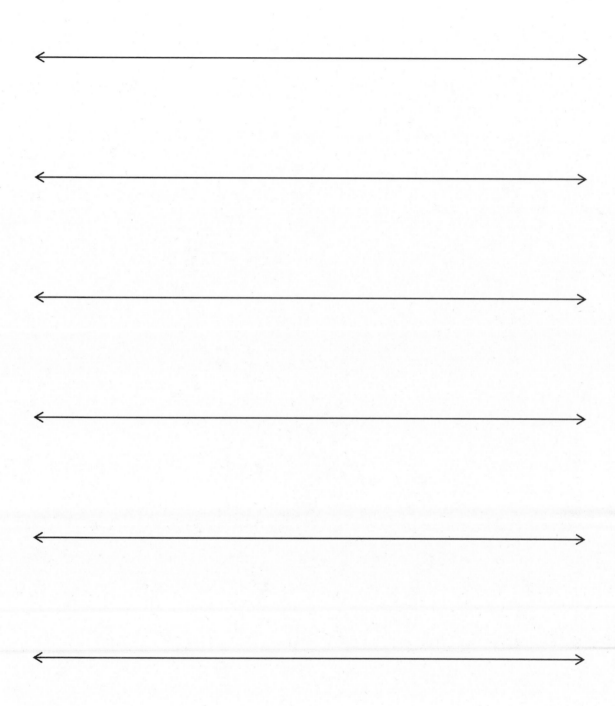

empty number line - from Lesson 8

Lesson 12:   Subtract fractions greater than or equal to 1.

93

© 2018 Great Minds®. eureka-math.org

Mark jogged $3\frac{5}{7}$ km. His sister jogged $2\frac{4}{5}$ km. How much farther did Mark jog than his sister?

Read   Draw   Write

Lesson 13:   Use fraction benchmark numbers to assess reasonableness of addition
                  and subtraction equations.

95

© 2018 Great Minds®. eureka-math.org

Name _____   Date _____

1. Are the following expressions greater than or less than 1? Circle the correct answer.

     a. $\frac{1}{2} + \frac{2}{7}$ greater than 1   less than 1

     b. $\frac{5}{8} + \frac{3}{5}$ greater than 1   less than 1

     c. $1\frac{1}{4} - \frac{1}{3}$ greater than 1   less than 1

     d. $3\frac{5}{8} - 2\frac{5}{9}$ greater than 1   less than 1

2. Are the following expressions greater than or less than $\frac{1}{2}$? Circle the correct answer.

     a. $\frac{1}{4} + \frac{2}{3}$ greater than $\frac{1}{2}$ less than $\frac{1}{2}$

     b. $\frac{3}{7} - \frac{1}{8}$ greater than $\frac{1}{2}$ less than $\frac{1}{2}$

     c. $1\frac{1}{7} - \frac{7}{8}$ greater than $\frac{1}{2}$ less than $\frac{1}{2}$

     d. $\frac{3}{7} + \frac{2}{6}$ greater than $\frac{1}{2}$ less than $\frac{1}{2}$

3. Use >, <, or = to make the following statements true.

     a. $5\frac{2}{3} + 3\frac{3}{4}$ _____ $8\frac{2}{3}$ b. $4\frac{5}{8} - 3\frac{2}{5}$ _____ $1\frac{5}{8} + \frac{2}{5}$

     c. $5\frac{1}{2} + 1\frac{3}{7}$ _____ $6 + \frac{13}{14}$ d. $15\frac{4}{7} - 11\frac{2}{5}$ _____ $4\frac{4}{7} + \frac{2}{5}$

EUREKA MATH   **Lesson 13:**   Use fraction benchmark numbers to assess reasonableness of addition   **97**
                                            and subtraction equations.

© 2018 Great Minds®. eureka-math.org

4. Is it true that $4\frac{3}{5} - 3\frac{2}{3} = 1 + \frac{3}{5} + \frac{2}{3}$? Prove your answer.

5. Jackson needs to be $1\frac{3}{4}$ inches taller in order to ride the roller coaster. Since he can't wait, he puts on a pair of boots that add $1\frac{1}{6}$ inches to his height and slips an insole inside to add another $\frac{1}{8}$ inches to his height. Will this make Jackson appear tall enough to ride the roller coaster?

6. A baker needs 5 lb of butter for a recipe. She found 2 portions that each weigh $1\frac{1}{6}$ lb and a portion that weighs $2\frac{2}{7}$ lb. Does she have enough butter for her recipe?

          Lesson 13:   Use fraction benchmark numbers to assess reasonableness of addition and subtraction equations.

EUREKA MATH

Name _____   Date _____

1. Circle the correct answer.

    a. $\frac{1}{2} + \frac{5}{12}$ greater than 1   less than 1

    b. $2\frac{7}{8} - 1\frac{7}{9}$ greater than 1   less than 1

    c. $1\frac{1}{12} - \frac{7}{10}$ greater than $\frac{1}{2}$ less than $\frac{1}{2}$

    d. $\frac{3}{7} + \frac{1}{8}$ greater than $\frac{1}{2}$ less than $\frac{1}{2}$

2. Use >, <, or = to make the following statement true.

$$4\frac{4}{5} + 3\frac{2}{3} \text{_____} 8\frac{1}{2}$$

EUREKA MATH®

Lesson 13:   Use fraction benchmark numbers to assess reasonableness of addition
and subtraction equations.

© 2018 Great Minds®. eureka-math.org

99

Problem 1

For a large order, Mr. Magoo made $\frac{3}{8}$ kg of fudge in his bakery. He then got $\frac{1}{6}$ kg from his sister's bakery. If he needs a total of $1\frac{1}{2}$ kg, how much more fudge does he need to make?

Problem 2

During lunch, Charlie drinks $2\frac{3}{4}$ cups of milk. Allison drinks $\frac{3}{8}$ cup of milk. Carmen drinks $1\frac{1}{6}$ cups of milk. How much milk do the 3 students drink?

Read **Draw** **Write**

Name _____   Date _____

1. Rearrange the terms so that you can add or subtract mentally. Then, solve.

   a. $\frac{1}{4} + 2\frac{2}{3} + \frac{7}{4} + \frac{1}{3}$

   b. $2\frac{3}{5} - \frac{3}{4} + \frac{2}{5}$

   c. $4\frac{3}{7} - \frac{3}{4} - 2\frac{1}{4} - \frac{3}{7}$

   d. $\frac{5}{6} + \frac{1}{3} - \frac{4}{3} + \frac{1}{6}$

2. Fill in the blank to make the statement true.

   a. $11\frac{2}{5} - 3\frac{2}{3} - \frac{11}{3} =$ _____

   b. $11\frac{7}{8} + 3\frac{1}{5} -$ _____ $= 15$

c. $\dfrac{5}{12} - \underline{\hspace{1cm}} + \dfrac{5}{4} = \dfrac{2}{3}$

d. $\underline{\hspace{1cm}} - 30 - 7\dfrac{1}{4} = 21\dfrac{2}{3}$

e. $\dfrac{24}{5} + \underline{\hspace{1cm}} + \dfrac{8}{7} = 9$

f. $11.1 + 3\dfrac{1}{10} - \underline{\hspace{1cm}} = \dfrac{99}{10}$

3.   DeAngelo needs 100 lb of garden soil to landscape a building. In the company's storage area, he finds 2 cases holding $24\dfrac{3}{4}$ lb of garden soil each, and a third case holding $19\dfrac{3}{8}$ lb. How much gardening soil does DeAngelo still need in order to do the job?

EUREKA
MATH

4. Volunteers helped clean up 8.2 kg of trash in one neighborhood and $11\frac{1}{2}$ kg in another. They sent $1\frac{1}{4}$ kg to be recycled and threw the rest away. How many kilograms of trash did they throw away?

Name _____ Date _____

Fill in the blank to make the statement true.

1. $1\frac{3}{4} + \frac{1}{6} +$ _____ $= 7\frac{1}{2}$

2. $8\frac{4}{5} - \frac{2}{3} -$ _____ $= 3\frac{1}{10}$

Name _____   Date _____

Solve the word problems using the RDW strategy. Show all of your work.

1. In a race, the-second place finisher crossed the finish line $1\frac{1}{3}$ minutes after the winner. The third-place finisher was $1\frac{3}{4}$ minutes behind the second-place finisher. The third-place finisher took $34\frac{2}{3}$ minutes. How long did the winner take?

2. John used $1\frac{3}{4}$ kg of salt to melt the ice on his sidewalk. He then used another $3\frac{4}{5}$ kg on the driveway. If he originally bought 10 kg of salt, how much does he have left?

Lesson 15:   Solve multi-step word problems; assess reasonableness of solutions using benchmark numbers.

© 2018 Great Minds®. eureka-math.org

109

3. Sinister Stan stole $3\frac{3}{4}$ oz of slime from Messy Molly, but his evil plans require $6\frac{3}{8}$ oz of slime. He stole another $2\frac{3}{5}$ oz of slime from Rude Ralph. How much more slime does Sinister Stan need for his evil plan?

4. Gavin had 20 minutes to do a three-problem quiz. He spent $9\frac{3}{4}$ minutes on Problem 1 and $3\frac{4}{5}$ minutes on Problem 2. How much time did he have left for Problem 3? Write the answer in minutes and seconds.

Lesson 15:   Solve multi-step word problems; assess reasonableness of solutions
                            using benchmark numbers.

          © 2018 Great Minds®. eureka-math.org

5. Matt wants to shave $2\frac{1}{2}$ minutes off his 5K race time. After a month of hard training, he managed to lower his overall time from $21\frac{1}{5}$ minutes to $19\frac{1}{4}$ minutes. By how many more minutes does Matt need to lower his race time?

Lesson 15: Solve multi-step word problems; assess reasonableness of solutions using benchmark numbers.

© 2018 Great Minds®. eureka-math.org

111

Name _____   Date _____

Solve the word problem using the RDW strategy. Show all of your work.

Cheryl bought a sandwich for $5\frac{1}{2}$ dollars and a drink for \$2.60. If she paid for her meal with a \$10 bill, how much money did she have left? Write your answer as a fraction and in dollars and cents.

Names _____ and _____   Date _____

1. Draw the following ribbons. When finished, compare your work to your partner's.

    a. 1 ribbon. The piece shown below is only $\frac{1}{3}$ of the whole. Complete the drawing to show the whole ribbon.

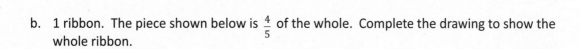

    b. 1 ribbon. The piece shown below is $\frac{4}{5}$ of the whole. Complete the drawing to show the whole ribbon.

    c. 2 ribbons, A and B. One third of A is equal to all of B. Draw a picture of the ribbons.

    d. 3 ribbons, C, D, and E. C is half the length of D. E is twice as long as D. Draw a picture of the ribbons.

2. Half of Robert's piece of wire is equal to $\frac{2}{3}$ of Maria's wire. The total length of their wires is 10 feet. How much longer is Robert's wire than Maria's?

3. Half of Sarah's wire is equal to $\frac{2}{5}$ of Daniel's. Chris has 3 times as much as Sarah. In all, their wire measures 6 ft. How long is Sarah's wire in feet?

EUREKA
MATH

Name _____   Date _____

Draw the following ribbons.

   a.   1 ribbon. The piece shown below is only $\frac{1}{3}$ of the whole. Complete the drawing to show the whole ribbon.

   b.   1 ribbon. The piece shown below is $\frac{1}{4}$ of the whole. Complete the drawing to show the whole ribbon.

   c.   3 ribbons, A, B, and C. 1 third of A is the same length as B. C is half as long as B. Draw a picture of the ribbons.

Grade 5
Module 4

The following line plot shows the growth, in inches, of 10 bean plants during their second week after sprouting:

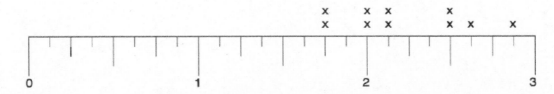

Bean Growth in Inches During Week Two

a. What is the measurement of the shortest plant?

b. How many plants measure $2\frac{1}{2}$ inches?

c. What is the measurement of the tallest plant?

d. What is the difference between the longest and shortest measurements?

Read   **Draw**   **Write**

Lesson 1:   Measure and compare pencil lengths to the nearest $\frac{1}{2}$, $\frac{1}{4}$, and $\frac{1}{8}$ of an inch, and analyze the data through line plots.

Name _____ Date _____

1. Estimate the length of your pencil to the nearest inch. _____

2. Using a ruler, measure your pencil strip to the nearest $\frac{1}{2}$ inch, and mark the measurement with an X above the ruler below. Construct a line plot of your classmates' pencil measurements.

3. Using a ruler, measure your pencil strip to the nearest $\frac{1}{4}$ inch, and mark the measurement with an X above the ruler below. Construct a line plot of your classmates' pencil measurements.

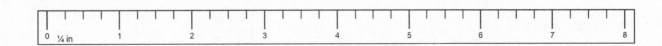

4. Using a ruler, measure your pencil strip to the nearest $\frac{1}{8}$ inch, and mark the measurement with an X above the ruler below. Construct a line plot of your classmates' pencil measurements.

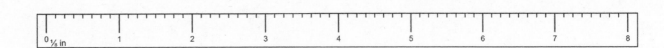

EUREKA MATH®

Lesson 1: Measure and compare pencil lengths to the nearest $\frac{1}{2}$, $\frac{1}{4}$, and $\frac{1}{8}$ of an inch, and analyze the data through line plots.

123

© 2018 Great Minds®. eureka-math.org

5. Use all three of your line plots to complete the following:

   a. Compare the three plots, and write one sentence that describes how the plots are alike and one sentence that describes how they are different.

   b. What is the difference between the measurements of the longest and shortest pencils on each of the three line plots?

   c. Write a sentence describing how you could create a more precise ruler to measure your pencil strip.

Lesson 1:   Measure and compare pencil lengths to the nearest $\frac{1}{2}$, $\frac{1}{4}$, and $\frac{1}{8}$ of an inch, and analyze the data through line plots.

© 2018 Great Minds®. eureka-math.org

EUREKA
MATH

Name _____   Date _____

1. Draw a line plot for the following data measured in inches:

$1\frac{1}{2}$, $2\frac{3}{4}$, 3, $2\frac{3}{4}$, $2\frac{1}{2}$, $2\frac{3}{4}$, $3\frac{3}{4}$, 3, $3\frac{1}{2}$, $2\frac{1}{2}$, $3\frac{1}{2}$

2. Explain how you decided to divide your wholes into fractional parts and how you decided where your number scale should begin and end.

EUREKA MATH

Lesson 1:   Measure and compare pencil lengths to the nearest $\frac{1}{2}$, $\frac{1}{4}$, and $\frac{1}{8}$ of an inch, and analyze the data through line plots.

© 2018 Great Minds®. eureka-math.org

125

The line plot shows the number of miles run by Noland in his PE class last month, which is rounded to the nearest quarter mile.

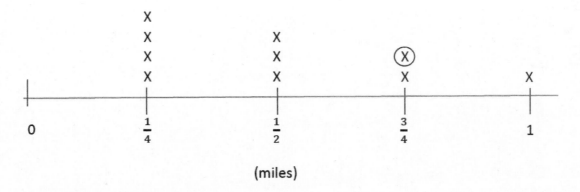

(miles)

a. If Noland ran once a day, how many days did he run?

b. How many miles did Noland run altogether last month?

c. Look at the circled data point. The actual distance Noland ran that day was at least _____ mile and less than _____ mile.

Read   **Draw**   **Write**

Name _____   Date _____

1. Draw a picture to show the division. Write a division expression using unit form. Then, express your answer as a fraction. The first one is partially done for you.

   a. $1 \div 5 = 5$ fifths $\div 5 = 1$ fifth $= \dfrac{1}{5}$

   b. $3 \div 4$

   c. $6 \div 4$

2. Draw to show how 2 children can equally share 3 cookies. Write an equation, and express your answer as a fraction.

3. Carly and Gina read the following problem in their math class:

   Seven cereal bars were shared equally by 3 children. How much did each child receive?

   Carly and Gina solve the problem differently. Carly gives each child 2 whole cereal bars and then divides the remaining cereal bar among the 3 children. Gina divides all the cereal bars into thirds and shares the thirds equally among the 3 children.

   a. Illustrate both girls' solutions.

   b. Explain why they are both right.

Lesson 2:   Interpret a fraction as division.

EUREKA
MATH

4. Fill in the blanks to make true number sentences.

   a. $2 \div 3 = $ _____

   b. $15 \div 8 = $ _____

   c. $11 \div 4 = $ _____

   d. $\dfrac{3}{2} = $ _____ $\div$ _____

   e. $\dfrac{9}{13} = $ _____ $\div$ _____

   f. $1\dfrac{1}{3} = $ _____ $\div$ _____

EUREKA
MATH

Name _____   Date _____

1. Draw a picture that shows the division expression. Then, write an equation and solve.

    a. $3 \div 9$ b. $4 \div 3$

2. Fill in the blanks to make true number sentences.

    a. $21 \div 8 = $ ___   b. $\dfrac{7}{4} = $ _____ $\div$ _____   c. $4 \div 9 = $ ___   d. $1\dfrac{2}{7} = $ _____ $\div$ _____

Hudson is choosing a seat in art class. He scans the room and sees a 4-person table with 1 bucket of art supplies, a 6-person table with 2 buckets of supplies, and a 5-person table with 2 buckets of supplies. Which table should Hudson choose if he wants the largest share of art supplies? Support your answer with pictures.

_____

_____

_____

_____

Read   **Draw**   **Write**

Name _____ Date _____

1. Fill in the chart. The first one is done for you.

Division Expression	Unit Forms	Improper Fraction	Mixed Numbers	Standard Algorithm (Write your answer in whole numbers and fractional units. Then check.)
a. $5 \div 4$	20 fourths ÷ 4 = 5 fourths	$\frac{5}{4}$	$1\frac{1}{4}$	$4\overline{)5}$ with $1\frac{1}{4}$ above, -4, 1. Check: $4 \times 1\frac{1}{4} = 1\frac{1}{4} + 1\frac{1}{4} + 1\frac{1}{4} + 1\frac{1}{4}$ $= 4 + \frac{4}{4}$ $= 4 + 1$ $= 5$
b. $3 \div 2$	___ halves ÷ 2 = ___ halves		$1\frac{1}{2}$	
c. ___ ÷ ___	24 fourths ÷ 4 = 6 fourths			$4\overline{)6}$
d. $5 \div 2$		$\frac{5}{2}$	$2\frac{1}{2}$	

2. A principal evenly distributes 6 reams of copy paper to 8 fifth-grade teachers.

   a. How many reams of paper does each fifth-grade teacher receive? Explain how you know using pictures, words, or numbers.

   b. If there were twice as many reams of paper and half as many teachers, how would the amount each teacher receives change? Explain how you know using pictures, words, or numbers.

3. A caterer has prepared 16 trays of hot food for an event. The trays are placed in warming boxes for delivery. Each box can hold 5 trays of food.

   a. How many warming boxes are necessary for delivery if the caterer wants to use as few boxes as possible? Explain how you know.

   b. If the caterer fills a box completely before filling the next box, what fraction of the last box will be empty?

Lesson 3: Interpret a fraction as division.

EUREKA MATH

Name _____   Date _____

A baker made 9 cupcakes, each a different type. Four people want to share them equally. How many cupcakes will each person get?

Fill in the chart to show how to solve the problem.

Division Expression	Unit Forms	Fractions and Mixed numbers	Standard Algorithm

Draw to show your thinking:

Four grade levels need equal time for indoor recess, and the gym is available for three hours.

a. How many hours of recess will each grade level receive? Draw a picture to support your answer.

b. How many minutes?

Read   **Draw**   **Write**

c. If the gym can accommodate two grade levels at once, how many hours of recess will 2 grade levels receive in 3 hours?

Read   **Draw**   **Write**

Lesson 4:   Use tape diagrams to model fractions as division.

EUREKA
MATH

Name _____   Date _____

1. Draw a tape diagram to solve. Express your answer as a fraction. Show the multiplication sentence to check your answer. The first one is done for you.

    a. $1 \div 3 = \frac{1}{3}$

    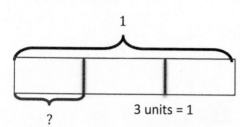

    3 units = 1

    1 unit = $1 \div 3$

    $= \frac{1}{3}$

    Check: $3 \times \frac{1}{3}$

    $= \frac{1}{3} + \frac{1}{3} + \frac{1}{3}$

    $= \frac{3}{3}$

    $= 1$

    b. $2 \div 3 = $ ___

    c. $7 \div 5 = $ ___

    d. $14 \div 5 = $ ___

EUREKA MATH

Lesson 4:   Use tape diagrams to model fractions as division.

143

© 2018 Great Minds®. eureka-math.org

2. Fill in the chart. The first one is done for you.

Division Expression	Fraction	Between which two whole numbers is your answer?	Standard Algorithm
a. $13 \div 3$	$\dfrac{13}{3}$	4 and 5	$4\frac{1}{3}$ $3\overline{)\begin{array}{l}13\\ -12\\ \hline 1\end{array}}$
b. $6 \div 7$		0 and 1	$7\overline{)6}$
c. ___ $\div$ ___	$\dfrac{55}{10}$		$\overline{)}$
d. ___ $\div$ ___	$\dfrac{32}{40}$		$40\overline{)32}$

Lesson 4:   Use tape diagrams to model fractions as division.

EUREKA MATH

3. Greg spent $4 on 5 packs of sport cards.

   a. How much did Greg spend on each pack?

   b. If Greg spent half as much money and bought twice as many packs of cards, how much did he spend on each pack? Explain your thinking.

4. Five pounds of birdseed is used to fill 4 identical bird feeders.

   a. What fraction of the birdseed will be needed to fill each feeder?

   b. How many pounds of birdseed are used to fill each feeder? Draw a tape diagram to show your thinking.

   c. How many ounces of birdseed are used to fill three bird feeders?

Name _____   Date _____

Matthew and his 3 siblings are weeding a flower bed with an area of 9 square yards. If they share the job equally, how many square yards of the flower bed will each child need to weed? Use a tape diagram to show your thinking.

Name _____   Date _____

1. A total of 2 yards of fabric is used to make 5 identical pillows. How much fabric is used for each pillow?

2. An ice cream shop uses 4 pints of ice cream to make 6 sundaes. How many pints of ice cream are used for each sundae?

3. An ice cream shop uses 6 bananas to make 4 identical sundaes. How many bananas are used in each sundae? Use a tape diagram to show your work.

Lesson 5:   Solve word problems involving the division of whole numbers with
answers in the form of fractions or whole numbers.

149

© 2018 Great Minds®. eureka-math.org

4. Julian has to read 4 articles for school. He has 8 nights to read them. He decides to read the same number of articles each night.

   a. How many articles will he have to read per night?

   b. What fraction of the reading assignment will he read each night?

5. 40 students shared 5 pizzas equally. How much pizza will each student receive? What fraction of the pizza did each student receive?

6. Lillian had 2 two-liter bottles of soda, which she distributed equally between 10 glasses.

   a. How much soda was in each glass? Express your answer as a fraction of a liter.

Lesson 5: Solve word problems involving the division of whole numbers with answers in the form of fractions or whole numbers.

© 2018 Great Minds®. eureka-math.org

EUREKA MATH

b. Express your answer as a decimal number of liters.

c. Express your answer as a whole number of milliliters.

7. The Calef family likes to paddle along the Susquehanna River.

   a. They paddled the same distance each day over the course of 3 days, traveling a total of 14 miles. How many miles did they travel each day? Show your thinking in a tape diagram.

   b. If the Calefs went half their daily distance each day but extended their trip to twice as many days, how far would they travel?

EUREKA MATH®

Lesson 5:   Solve word problems involving the division of whole numbers with answers in the form of fractions or whole numbers.

© 2018 Great Minds®. eureka-math.org

151

Name _____   Date _____

A grasshopper covered a distance of 5 yards in 9 equal hops. How many yards did the grasshopper travel on each hop?

a. Draw a picture to support your work.

b. How many yards did the grasshopper travel after hopping twice?

Lesson 5:   Solve word problems involving the division of whole numbers with   **153**
            answers in the form of fractions or whole numbers.

© 2018 Great Minds®. eureka-math.org

Olivia is half the age of her brother, Adam. Olivia's sister, Ava, is twice as old as Adam. Adam is 4 years old. How old is each sibling? Use tape diagrams to show your thinking.

_____

_____

_____

_____

Read   **Draw**   **Write**

Name _____ Date _____

1. Find the value of each of the following.

a.

$\frac{1}{3}$ of 9 =

$\frac{2}{3}$ of 9 =

$\frac{3}{3}$ of 9 =

b.

$\frac{1}{3}$ of 15 =

$\frac{2}{3}$ of 15 =

$\frac{3}{3}$ of 15 =

c.

$\frac{1}{5}$ of 20 =

$\frac{4}{5}$ of 20 =

$\frac{}{5}$ of 20 = 20

d.

$\frac{1}{8}$ of 24 =

$\frac{6}{8}$ of 24 =

$\frac{3}{8}$ of 24 =

$\frac{7}{8}$ of 24 =

$\frac{4}{8}$ of 24 =

EUREKA
MATH®

Lesson 6:   Relate fractions as division to fraction of a set.

157

2. Find $\frac{4}{7}$ of 14. Draw a set, and shade to show your thinking.

3. How does knowing $\frac{1}{8}$ of 24 help you find three-eighths of 24? Draw a picture to explain your thinking.

4. There are 32 students in a class. Of the class, $\frac{3}{8}$ of the students bring their own lunches. How many students bring their lunches?

5. Jack collected 18 ten-dollar bills while selling tickets for a show. He gave $\frac{1}{6}$ of the bills to the theater and kept the rest. How much money did he keep?

Lesson 6:   Relate fractions as division to fraction of a set.

EUREKA
MATH

Name _____   Date _____

1. Find the value of each of the following.

a. $\dfrac{1}{4}$ of 16 =

b. $\dfrac{3}{4}$ of 16 =

2. Out of 18 cookies, $\dfrac{2}{3}$ are chocolate chip. How many of the cookies are chocolate chip?

Lesson 6:   Relate fractions as division to fraction of a set.

159

© 2018 Great Minds®. eureka-math.org

Mr. Peterson bought a case (24 boxes) of fruit juice. One-third of the drinks were grape, and two-thirds were cranberry. How many boxes of each flavor did Mr. Peterson buy? Show your work using a tape diagram or an array.

Read   **Draw**   **Write**

Lesson 7:   Multiply any whole number by a fraction using tape diagrams.

161

© 2018 Great Minds®. eureka-math.org

Name _____   Date _____

1. Solve using a tape diagram.

   a. $\frac{1}{3}$ of 18   b. $\frac{1}{2}$ of 36

   c. $\frac{3}{4} \times 24$ d. $\frac{3}{8} \times 24$

   e. $\frac{4}{5} \times 25$ f. $\frac{1}{7} \times 140$

   g. $\frac{1}{4} \times 9$ h. $\frac{2}{5} \times 12$

   i. $\frac{2}{3}$ of a number is 10. What's the number?   j. $\frac{3}{4}$ of a number is 24. What's the number?

2. Solve using tape diagrams.

   a. There are 48 students going on a field trip. One-fourth are girls. How many boys are going on the trip?

   b. Three angles are labeled below with arcs. The smallest angle is $\frac{3}{8}$ as large as the 160° angle. Find the value of angle a.

   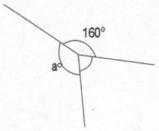

   c. Abbie spent $\frac{5}{8}$ of her money and saved the rest. If she spent $45, how much money did she have at first?

   d. Mrs. Harrison used 16 ounces of dark chocolate while baking. She used $\frac{2}{5}$ of the chocolate to make some frosting and used the rest to make brownies. How much more chocolate did Mrs. Harrison use in the brownies than in the frosting?

Lesson 7:   Multiply any whole number by a fraction using tape diagrams.

EUREKA
MATH

Name _____   Date _____

Solve using a tape diagram.

a. $\frac{3}{5}$ of 30

b. $\frac{3}{5}$ of a number is 30. What's the number?

c. Mrs. Johnson baked 2 dozen cookies. Two-thirds of the cookies were oatmeal. How many oatmeal cookies did Mrs. Johnson bake?

Grade 5 Mathematics Reference Sheet

FORMULAS

Right Rectangular Prism

Volume = lwh

Volume = Bh

CONVERSIONS

1 centimeter = 10 millimeters

1 meter = 100 centimeters = 1,000 millimeters

1 kilometer = 1,000 meters

1 gram = 1,000 milligrams

1 kilogram = 1,000 grams

1 pound = 16 ounces

1 ton = 2,000 pounds

1 cup = 8 fluid ounces

1 pint = 2 cups

1 quart = 2 pints

1 gallon = 4 quarts

1 liter = 1,000 milliliters

1 kiloliter = 1,000 liters

1 mile = 5,280 feet

1 mile = 1,760 yards

Sasha organizes the art gallery in her town's community center. This month, she has 24 new pieces to add to the gallery. Of the new pieces, $\frac{1}{6}$ of them are photographs, and $\frac{2}{3}$ of them are paintings. How many more paintings are there than photos?

Read   **Draw**   **Write**

Name _____   Date _____

1. Laura and Sean find the product of $\frac{2}{3} \times 4$ using different methods.

   Laura: It's 2 thirds of 4.   *Sean:* It's 4 groups of 2 thirds.

   $$\frac{2}{3} \times 4 = \frac{4}{3} + \frac{4}{3} = 2 \times \frac{4}{3} = \frac{8}{3} \qquad \frac{2}{3} + \frac{2}{3} + \frac{2}{3} + \frac{2}{3} = 4 \times \frac{2}{3} = \frac{8}{3}$$

   Use words, pictures, or numbers to compare their methods in the space below.

2. Rewrite the following addition expressions as fractions as shown in the example.

   Example: $\frac{2}{3} + \frac{2}{3} + \frac{2}{3} + \frac{2}{3} = \frac{4 \times 2}{3} = \frac{8}{3}$

   a. $\frac{7}{4} + \frac{7}{4} + \frac{7}{4} =$ b. $\frac{14}{5} + \frac{14}{5} =$ c. $\frac{4}{7} + \frac{4}{7} + \frac{4}{7} =$

3. Solve and model each problem as a fraction of a set and as repeated addition.

   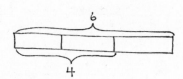   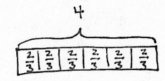

   Example: $\frac{2}{3} \times 6 = 2 \times \frac{6}{3} = 2 \times 2 = 4$ $6 \times \frac{2}{3} = \frac{6 \times 2}{3} = 4$

   a. $\frac{1}{2} \times 8$ $8 \times \frac{1}{2}$

   b. $\frac{3}{5} \times 10$ $10 \times \frac{3}{5}$

EUREKA MATH   **Lesson 8:**   Relate a fraction of a set to the repeated addition interpretation of   **171**
fraction multiplication.

© 2018 Great Minds®. eureka-math.org

4. Solve each problem in two different ways as modeled in the example.

Example: $6 \times \dfrac{2}{3} = \dfrac{6 \times 2}{3} = \dfrac{3 \times 2 \times 2}{3} = \dfrac{3 \times 4}{3} = 4$ $6 \times \dfrac{2}{3} = \dfrac{\overset{2}{\cancel{6}} \times 2}{\underset{1}{\cancel{3}}} = 4$

a. $14 \times \dfrac{3}{7}$ $14 \times \dfrac{3}{7}$

b. $\dfrac{3}{4} \times 36$ $\dfrac{3}{4} \times 36$

c. $30 \times \dfrac{13}{10}$ $30 \times \dfrac{13}{10}$

d. $\dfrac{9}{8} \times 32$ $\dfrac{9}{8} \times 32$

5. Solve each problem any way you choose.

a. $\dfrac{1}{2} \times 60$ $\dfrac{1}{2}$ minute = _____ seconds

b. $\dfrac{3}{4} \times 60$ $\dfrac{3}{4}$ hour = _____ minutes

c. $\dfrac{3}{10} \times 1,000$ $\dfrac{3}{10}$ kilogram = _____ grams

d. $\dfrac{4}{5} \times 100$ $\dfrac{4}{5}$ meter = _____ centimeters

EUREKA
MATH

Name _____   Date _____

Solve each problem in two different ways as modeled in the example.

Example: $\dfrac{2}{3} \times 6 = \dfrac{2 \times 6}{3} = \dfrac{12}{3} = 4$ $\dfrac{2}{3} \times 6 = \dfrac{2 \times \cancel{6}^{2}}{\cancel{3}_{1}} = 4$

a. $\dfrac{2}{3} \times 15$ $\dfrac{2}{3} \times 15$

b. $\dfrac{5}{4} \times 12$ $\dfrac{5}{4} \times 12$

EUREKA MATH®

Lesson 8: Relate a fraction of a set to the repeated addition interpretation of fraction multiplication.

© 2018 Great Minds®. eureka-math.org

173

There are 42 people at a museum. Two-thirds of them are children. How many children are at the museum?

Extension: If 13 of the children are girls, how many more boys than girls are at the museum?

_____

_____

_____

_____

Read   **Draw**   **Write**

Name _____ Date _____

1. Convert. Show your work using a tape diagram or an equation. The first one is done for you.

a. $\frac{1}{2}$ yard = ___1$\frac{1}{2}$___ feet $\frac{1}{2}$ yard = $\frac{1}{2}$ × 1 yard $\quad$ = $\frac{1}{2}$ × 3 feet $\quad$ = $\frac{3}{2}$ feet $\quad$ = $1\frac{1}{2}$ feet	b. $\frac{1}{3}$ foot = _____ inches $\frac{1}{3}$ foot = $\frac{1}{3}$ × 1 foot $\quad$ = $\frac{1}{3}$ × 12 inches $\quad$ = 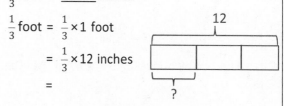
c. $\frac{5}{6}$ year = _____ months	d. $\frac{4}{5}$ meter = _____ centimeters
e. $\frac{2}{3}$ hour = _____ minutes	f. $\frac{3}{4}$ yard = _____ inches

Lesson 9:   Find a fraction of a measurement, and solve word problems.

177

2. Mrs. Lang told her class that the class's pet hamster is $\frac{1}{4}$ ft in length. How long is the hamster in inches?

3. At the market, Mr. Paul bought $\frac{7}{8}$ lb of cashews and $\frac{3}{4}$ lb of walnuts.

   a. How many ounces of cashews did Mr. Paul buy?

   b. How many ounces of walnuts did Mr. Paul buy?

   c. How many more ounces of cashews than walnuts did Mr. Paul buy?

   d. If Mrs. Toombs bought $1\frac{1}{2}$ pounds of pistachios, who bought more nuts, Mr. Paul or Mrs. Toombs? How many ounces more?

4. A jewelry maker purchased 20 inches of gold chain. She used $\frac{3}{8}$ of the chain for a bracelet. How many inches of gold chain did she have left?

       Lesson 9:   Find a fraction of a measurement, and solve word problems.

EUREKA
MATH

Name _____ Date _____

1. Express 36 minutes as a fraction of an hour: 36 minutes = _____ hour

2. Solve.

   a. $\frac{2}{3}$ feet = _____ inches   b. $\frac{2}{5}$ m = _____ cm   c. $\frac{5}{6}$ year = _____ months

EUREKA
MATH®

Lesson 9:   Find a fraction of a measurement, and solve word problems.

179

© 2018 Great Minds®. eureka-math.org

Bridget has \$240. She spent $\frac{3}{5}$ of her money and saved the rest. How much more money did she spend than save?

Read   **Draw**   **Write**

Name _____ Date _____

1. Write expressions to match the diagrams. Then, evaluate.

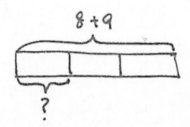

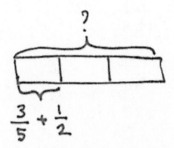

2. Write an expression to match, and then evaluate.

a. $\frac{1}{6}$ the sum of 16 and 20

b. Subtract 5 from $\frac{1}{3}$ of 23.

c. 3 times as much as the sum of $\frac{3}{4}$ and $\frac{2}{6}$

d. $\frac{2}{5}$ of the product of $\frac{5}{6}$ and 42

e. 8 copies of the sum of 4 thirds and 2 more

f. 4 times as much as 1 third of 8

3. Circle the expression(s) that give the same product as $\frac{4}{5} \times 7$. Explain how you know.

   $4 \div (7 \times 5)$ $7 \div 5 \times 4$ $(4 \times 7) \div 5$ $4 \div (5 \times 7)$ $4 \times \frac{7}{5}$ $7 \times \frac{4}{5}$

4. Use <, >, or = to make true number sentences without calculating. Explain your thinking.

   a. $4 \times 2 + 4 \times \frac{2}{3}$ $\bigcirc$ $3 \times \frac{2}{3}$

   b. $\left(5 \times \frac{3}{4}\right) \times \frac{2}{5}$ $\bigcirc$ $\left(5 \times \frac{3}{4}\right) \times \frac{2}{7}$

   c. $3 \times \left(3 + \frac{15}{12}\right)$ $\bigcirc$ $(3 \times 3) + \frac{15}{12}$

Lesson 10:   Compare and evaluate expressions with parentheses.

EUREKA
MATH

5. Collette bought milk for herself each month and recorded the amount in the table below. For (a)–(c), write an expression that records the calculation described. Then, solve to find the missing data in the table.

a. She bought $\frac{1}{4}$ of July's total in June.

b. She bought $\frac{3}{4}$ as much in September as she did in January and July combined.

Month	Amount (in gallons)
January	3
February	2
March	$1\frac{1}{4}$
April	
May	$\frac{7}{4}$
June	
July	2
August	1
September	
October	$\frac{1}{4}$

c. In April, she bought $\frac{1}{2}$ gallon less than twice as much as she bought in August.

d. Display the data from the table in a line plot.

e. How many gallons of milk did Collette buy from January to October?

Lesson 10:   Compare and evaluate expressions with parentheses.

Name _____   Date _____

1. Rewrite these expressions using words.

   a. $\frac{3}{4} \times \left(2\frac{2}{5} - \frac{5}{6}\right)$

   b. $2\frac{1}{4} + \frac{8}{3}$

3. Write an expression, and then solve.

   Three less than one-fourth of the product of eight thirds and nine

Name _____ Date _____

1. Kim and Courtney share a 16-ounce box of cereal. By the end of the week, Kim has eaten $\frac{3}{8}$ of the box, and Courtney has eaten $\frac{1}{4}$ of the box of cereal. What fraction of the box is left?

2. Mathilde has 20 pints of green paint. She uses $\frac{2}{5}$ of it to paint a landscape and $\frac{3}{10}$ of it while painting a clover. She decides that, for her next painting, she will need 14 pints of green paint. How much more paint will she need to buy?

Lesson 11:   Solve and create fraction word problems involving addition,
              subtraction, and multiplication.

© 2018 Great Minds®. eureka-math.org

189

3. Jack, Jill, and Bill each carried a 48-ounce bucket full of water down the hill. By the time they reached the bottom, Jack's bucket was only $\frac{3}{4}$ full, Jill's was $\frac{2}{3}$ full, and Bill's was $\frac{1}{6}$ full. How much water did they spill altogether on their way down the hill?

4. Mrs. Diaz makes 5 dozen cookies for her class. One-ninth of her 27 students are absent the day she brings the cookies. If she shares the cookies equally among the students who are present, how many cookies will each student get?

5. Create a story problem about a fish tank for the tape diagram below. Your story must include a fraction.

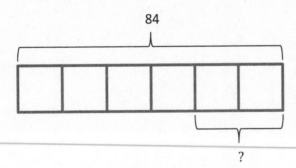

Lesson 11: Solve and create fraction word problems involving addition, subtraction, and multiplication.

EUREKA MATH

Name _____   Date _____

Use a tape diagram to solve.

$\frac{2}{3}$ of 5

Lesson 11:   Solve and create fraction word problems involving addition,
               subtraction, and multiplication.

© 2018 Great Minds®. eureka-math.org

191

Complete the table.

$\frac{2}{3}$ yard	_____ foot(feet)
4 pounds	_____ ounce(s)
8 tons	_____ pound(s)
$\frac{3}{4}$ gallon	_____ quart(s)
$\frac{5}{12}$ year	_____ month(s)
$\frac{4}{5}$ hour	_____ minute(s)

Read   **Draw**   **Write**

Lesson 12:   Solve and create fraction word problems involving addition, subtraction, and multiplication.

193

© 2018 Great Minds®. eureka-math.org

Name _____ Date _____

1. A baseball team played 32 games and lost 8. Katy was the catcher in $\frac{5}{8}$ of the winning games and $\frac{1}{4}$ of the losing games.

   a. What fraction of the games did the team win?

   b. In how many games did Katy play catcher?

2. In Mrs. Elliott's garden, $\frac{1}{8}$ of the flowers are red, $\frac{1}{4}$ of them are purple, and $\frac{1}{5}$ of the remaining flowers are pink. If there are 128 flowers, how many flowers are pink?

Lesson 12:   Solve and create fraction word problems involving addition,
subtraction, and multiplication.

195

© 2018 Great Minds®. eureka-math.org

3. Lillian and Darlene plan to get their homework finished within one hour. Darlene completes her math homework in $\frac{3}{5}$ hour. Lillian completes her math homework with $\frac{5}{6}$ hour remaining. Who completes her homework faster, and by how many minutes?

    Bonus: Give the answer as a fraction of an hour.

4. Create and solve a story problem about a baker and some flour whose solution is given by the expression $\frac{1}{4} \times (3 + 5)$.

Solve and create fraction word problems involving addition, subtraction, and multiplication.

© 2018 Great Minds®. eureka-math.org

EUREKA
MATH

5. Create and solve a story problem about a baker and 36 kilograms of an ingredient that is modeled by the following tape diagram. Include at least one fraction in your story.

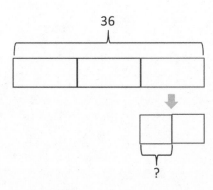

6. Of the students in Mr. Smith's fifth-grade class, $\frac{1}{3}$ were absent on Monday. Of the students in Mrs. Jacobs' class, $\frac{2}{5}$ were absent on Monday. If there were 4 students absent in each class on Monday, how many students are in each class?

Lesson 12: Solve and create fraction word problems involving addition, subtraction, and multiplication.

197

© 2018 Great Minds®. eureka-math.org

Name _____   Date _____

In a classroom, $\frac{1}{6}$ of the students are wearing blue shirts, and $\frac{2}{3}$ are wearing white shirts. There are 36 students in the class. How many students are wearing a shirt other than blue or white?

Lesson 12:   Solve and create fraction word problems involving addition,   199
              subtraction, and multiplication.

© 2018 Great Minds®. eureka-math.org

Name _____   Date _____

1. Solve. Draw a rectangular fraction model to show your thinking. Then, write a multiplication sentence. The first one has been done for you.

   a. Half of $\frac{1}{4}$ pan of brownies = ___$\frac{1}{8}$___ pan of brownies.

      $\frac{1}{2} \times \frac{1}{4} = \frac{1}{8}$

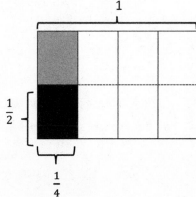

   b. Half of $\frac{1}{3}$ pan of brownies = _____ pan of brownies.

   c. A fourth of $\frac{1}{3}$ pan of brownies = _____ pan of brownies.

   d. $\frac{1}{4}$ of $\frac{1}{4}$

   e. $\frac{1}{2}$ of $\frac{1}{6}$

2. Draw rectangular fraction models of $3 \times \frac{1}{4}$ and $\frac{1}{3} \times \frac{1}{4}$. Compare multiplying a number by 3 and by 1 third.

3. $\frac{1}{2}$ of Ila's workspace is covered in paper. $\frac{1}{3}$ of the paper is covered in yellow sticky notes. What fraction of Ila's workspace is covered in yellow sticky notes? Draw a picture to support your answer.

4. A marching band is rehearsing in rectangular formation. $\frac{1}{5}$ of the marching band members play percussion instruments. $\frac{1}{2}$ of the percussionists play the snare drum. What fraction of all the band members play the snare drum?

5. Marie is designing a bedspread for her grandson's new bedroom. $\frac{2}{3}$ of the bedspread is covered in race cars, and the rest is striped. $\frac{1}{4}$ of the stripes are red. What fraction of the bedspread is covered in red stripes?

Lesson 13:   Multiply unit fractions by unit fractions.

EUREKA MATH

Name _____   Date _____

1. Solve. Draw a rectangular fraction model, and write a number sentence to show your thinking.

    $\frac{1}{3} \times \frac{1}{3} =$

2. Ms. Sheppard cuts $\frac{1}{2}$ of a piece of construction paper. She uses $\frac{1}{6}$ of the piece to make a flower. What fraction of the sheet of paper does she use to make the flower?

Solve by drawing a rectangular fraction model and writing a multiplication sentence. Beth had $\frac{1}{4}$ box of candy. She ate $\frac{1}{2}$ of the candy. What fraction of the whole box does she have left?

Extension: If Beth decides to refill the box, what fraction of the box would need to be refilled?

Read   **Draw**   **Write**

Name _____ Date _____

1. Solve. Draw a rectangular fraction model to explain your thinking. Then, write a number sentence. An example has been done for you.

   Example:

   $\frac{1}{2}$ of $\frac{2}{5}$ = $\frac{1}{2}$ of 2 fifths = 1 fifth(s)

   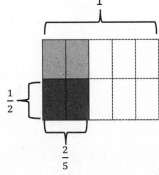

   $\frac{1}{2} \times \frac{2}{5} = \frac{2}{10} = \frac{1}{5}$

   a. $\frac{1}{3}$ of $\frac{3}{4} = \frac{1}{3}$ of _____ fourth(s) = _____ fourth(s)

   b. $\frac{1}{2}$ of $\frac{4}{5} = \frac{1}{2}$ of _____ fifth(s) = _____ fifth(s)

   c. $\frac{1}{2}$ of $\frac{2}{2}$ =

   d. $\frac{2}{3}$ of $\frac{1}{2}$ =

   e. $\frac{1}{2} \times \frac{3}{5}$ =

   f. $\frac{2}{3} \times \frac{1}{4}$ =

2. $\frac{5}{8}$ of the songs on Harrison's music player are hip-hop. $\frac{1}{3}$ of the remaining songs are rhythm and blues. What fraction of all the songs are rhythm and blues? Use a tape diagram to solve.

3. Three-fifths of the students in a room are girls. One-third of the girls have blond hair. One-half of the boys have brown hair.

a. What fraction of all the students are girls with blond hair?

b. What fraction of all the students are boys without brown hair?

4. Cody and Sam mowed the yard on Saturday. Dad told Cody to mow $\frac{1}{4}$ of the yard. He told Sam to mow $\frac{1}{3}$ of the remainder of the yard. Dad paid each of the boys an equal amount. Sam said, "Dad, that's not fair! I had to mow one-third, and Cody only mowed one-fourth!" Explain to Sam the error in his thinking. Draw a picture to support your reasoning.

EUREKA MATH

Name _____ Date _____

1. Solve. Draw a rectangular fraction model to explain your thinking. Then, write a number sentence.

    $\frac{1}{3}$ of $\frac{3}{7}$ =

2. In a cookie jar, $\frac{1}{4}$ of the cookies are chocolate chip, and $\frac{1}{2}$ of the rest are peanut butter. What fraction of all the cookies is peanut butter?

Kendra spent $\frac{1}{3}$ of her allowance on a book and $\frac{2}{5}$ on a snack. If she had four dollars remaining after purchasing a book and snack, what was the total amount of her allowance?

Read   Draw   Write

Name _____   Date _____

1. Solve. Draw a rectangular fraction model to explain your thinking. Then, write a multiplication sentence. The first one is done for you.

   a. $\frac{2}{3}$ of $\frac{3}{5}$

   $\frac{2}{3} \times \frac{3}{5} = \frac{6}{15} = \frac{2}{5}$

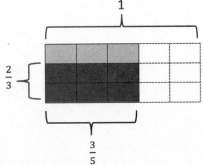

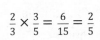

   b. $\frac{3}{5}$ of $\frac{4}{5} =$

   c. $\frac{2}{5}$ of $\frac{2}{3} =$

   d. $\frac{4}{5} \times \frac{2}{3} =$

   e. $\frac{3}{4} \times \frac{2}{3} =$

2. Multiply. Draw a rectangular fraction model if it helps you, or use the method in the example.

   Example: $\frac{6}{7} \times \frac{5}{8} = \frac{\overset{3}{\cancel{6}} \times 5}{7 \times \underset{4}{\cancel{8}}} = \frac{15}{28}$

   a. $\frac{3}{4} \times \frac{5}{6}$

   b. $\frac{4}{5} \times \frac{5}{8}$

EUREKA MATH

Lesson 15: Multiply non-unit fractions by non-unit fractions.

213

c. $\frac{2}{3} \times \frac{6}{7}$ d. $\frac{4}{9} \times \frac{3}{10}$

3. Phillip's family traveled $\frac{3}{10}$ of the distance to his grandmother's house on Saturday. They traveled $\frac{4}{7}$ of the remaining distance on Sunday. What fraction of the total distance to his grandmother's house was traveled on Sunday?

4. Santino bought a $\frac{3}{4}$-pound bag of chocolate chips. He used $\frac{2}{3}$ of the bag while baking. How many pounds of chocolate chips did he use while baking?

5. Farmer Dave harvested his corn. He stored $\frac{5}{9}$ of his corn in one large silo and $\frac{3}{4}$ of the remaining corn in a small silo. The rest was taken to market to be sold.

    a. What fraction of the corn was stored in the small silo?

    b. If he harvested 18 tons of corn, how many tons did he take to market?

EUREKA
MATH

Name _____   Date _____

1. Solve. Draw a rectangular fraction model to explain your thinking. Then, write a multiplication sentence.

a. $\frac{2}{3}$ of $\frac{3}{5}$ =

b. $\frac{4}{9} \times \frac{3}{8}$ =

2. A newspaper's cover page is $\frac{3}{8}$ text, and photographs fill the rest. If $\frac{2}{5}$ of the text is an article about endangered species, what fraction of the cover page is the article about endangered species?

Lesson 15:   Multiply non-unit fractions by non-unit fractions.

215

© 2018 Great Minds®. eureka-math.org

Name _____   Date _____

Solve and show your thinking with a tape diagram.

1. Mrs. Onusko made 60 cookies for a bake sale. She sold $\frac{2}{3}$ of them and gave $\frac{3}{4}$ of the remaining cookies to the students working at the sale. How many cookies did she have left?

2. Joakim is icing 30 cupcakes. He spreads mint icing on $\frac{1}{5}$ of the cupcakes and chocolate on $\frac{1}{2}$ of the remaining cupcakes. The rest will get vanilla icing. How many cupcakes have vanilla icing?

3. The Booster Club sells 240 cheeseburgers. $\frac{1}{4}$ of the cheeseburgers had pickles, $\frac{1}{2}$ of the remaining burgers had onions, and the rest had tomato. How many cheeseburgers had tomato?

4. DeSean is sorting his rock collection. $\frac{2}{3}$ of the rocks are metamorphic, and $\frac{3}{4}$ of the remainder are igneous rocks. If the 3 rocks left over are sedimentary, how many rocks does DeSean have?

5. Milan puts $\frac{1}{4}$ of her lawn-mowing money in savings and uses $\frac{1}{2}$ of the remaining money to pay back her sister. If she has $15 left, how much did she have at first?

6. Parks is wearing several rubber bracelets. $\frac{1}{3}$ of the bracelets are tie-dye, $\frac{1}{6}$ are blue, and $\frac{1}{3}$ of the remainder are camouflage. If Parks wears 2 camouflage bracelets, how many bracelets does he have on?

7. Ahmed spent $\frac{1}{3}$ of his money on a burrito and a water bottle. The burrito cost 2 times as much as the water. The burrito cost $4. How much money does Ahmed have left?

Lesson 16:   Solve word problems using tape diagrams and fraction-by-fraction multiplication.

© 2018 Great Minds®. eureka-math.org

EUREKA
MATH

Name _____   Date _____

Solve and show your thinking with a tape diagram.

Three-quarters of the boats in the marina are white, $\frac{4}{7}$ of the remaining boats are blue, and the rest are red. If there are 9 red boats, how many boats are in the marina?

Lesson 16:   Solve word problems using tape diagrams and fraction-by-fraction
              multiplication.

© 2018 Great Minds®. eureka-math.org

219

Ms. Casey grades 4 tests during her lunch. She grades $\frac{1}{3}$ of the remainder after school. If she still has 16 tests to grade after school, how many tests are there?

Read   **Draw**   **Write**

Name _____ Date _____

1. Multiply and model. Rewrite each expression as a multiplication sentence with decimal factors. The first one is done for you.

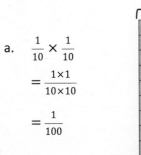

a. $\frac{1}{10} \times \frac{1}{10}$

$= \frac{1 \times 1}{10 \times 10}$

$= \frac{1}{100}$

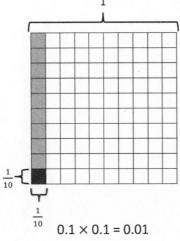

$0.1 \times 0.1 = 0.01$

b. $\frac{4}{10} \times \frac{3}{10}$

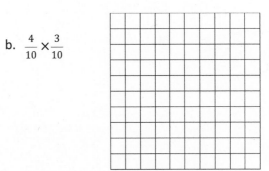

c. $\frac{1}{10} \times 1.4$

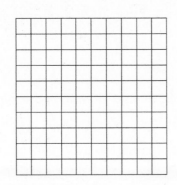   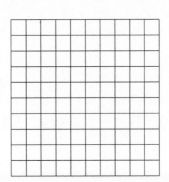

d. $\frac{6}{10} \times 1.7$

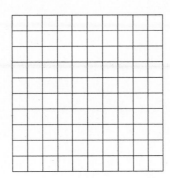   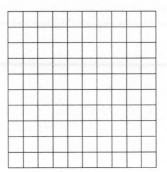

EUREKA
MATH®

2. Multiply. The first few are started for you.

a. $5 \times 0.7 =$ _____

$= 5 \times \dfrac{7}{10}$

$= \dfrac{5 \times 7}{10}$

$= \dfrac{35}{10}$

$= 3.5$

b. $0.5 \times 0.7 =$ _____

$= \dfrac{5}{10} \times \dfrac{7}{10}$

$= \dfrac{5 \times 7}{10 \times 10}$

$=$

c. $0.05 \times 0.7 =$ _____

$= \dfrac{5}{100} \times \dfrac{7}{10}$

$= \dfrac{\_\_ \times \_\_}{100 \times 10}$

$=$

d. $6 \times 0.3 =$ _____

e. $0.6 \times 0.3 =$ _____

f. $0.06 \times 0.3 =$ _____

g. $1.2 \times 4 =$ _____

h. $1.2 \times 0.4 =$ _____

i. $0.12 \times 0.4 =$ _____

3. A Boy Scout has a length of rope measuring 0.7 meter. He uses 2 tenths of the rope to tie a knot at one end. How many meters of rope are in the knot?

4. After just 4 tenths of a 2.5-mile race was completed, Lenox took the lead and remained there until the end of the race.

a. How many miles did Lenox lead the race?

b. Reid, the second-place finisher, developed a cramp with 3 tenths of the race remaining. How many miles did Reid run without a cramp?

Lesson 17:   Relate decimal and fraction multiplication.

EUREKA MATH

Name _____ Date _____

1. Multiply and model. Rewrite the expression as a number sentence with decimal factors.

$\frac{1}{10} \times 1.2$

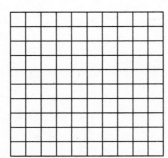
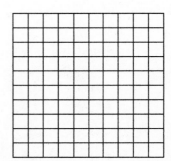

2. Multiply.

a. $1.5 \times 3 =$ _____

b. $1.5 \times 0.3 =$ _____

c. $0.15 \times 0.3 =$ _____

1,000,000	100,000	10,000	1,000	100	10	1	•	$\frac{1}{10}$	$\frac{1}{100}$	$\frac{1}{1000}$
Millions	Hundred Thousands	Ten Thousands	Thousands	Hundreds	Tens	Ones	•	Tenths	Hundredths	Thousandths
							•			
							•			
							•			
							•			
							•			
							•			
							•			
							•			
							•			
							•			

millions through thousandths place value chart

An adult female gorilla is 1.4 meters tall when standing upright. Her daughter is 3 tenths as tall.

How much more will the young female gorilla need to grow before she is as tall as her mother?

Read   **Draw**   **Write**

Name _____   Date _____

1. Multiply using both fraction form and unit form. Check your answer by counting the decimal places. The first one is done for you.

   a. $2.3 \times 1.8 = \frac{23}{10} \times \frac{18}{10}$

   $$= \frac{23 \times 18}{100}$$

   $$= \frac{414}{100}$$

   $$= 4.14$$

   ```
       2 3  tenths
   ×   1 8  tenths
       1 8 4
   +   2 3 0
       4 1 4  hundredths
   ```

   b. $2.3 \times 0.9 =$

   ```
       2 3  tenths
   ×      9  tenths
   ```

   c. $6.6 \times 2.8 =$

   d. $3.3 \times 1.4 =$

2. Multiply using fraction form and unit form. Check your answer by counting the decimal places. The first one is done for you.

   a. $2.38 \times 1.8 = \frac{238}{100} \times \frac{18}{10}$

   $$= \frac{238 \times 18}{1,000}$$

   $$= \frac{4,284}{1,000}$$

   $$= 4.284$$

   ```
       2 3 8  hundredths
   ×     1 8  tenths
       1 9 0 4
   +   2 3 8 0
     4,2 8 4  thousandths
   ```

   b. $2.37 \times 0.9 =$

   ```
       2 3 7  hundredths
   ×        9  tenths
   ```

   c. $6.06 \times 2.8 =$

   d. $3.3 \times 0.14 =$

3. Solve using the standard algorithm. Show your thinking about the units of your product. The first one is done for you.

a. $3.2 \times 0.6 = 1.92$

```
  3 2 tenths
×     6 tenths
  1 9 2 hundredths
```

$$\frac{32}{10} \times \frac{6}{10} = \frac{32 \times 6}{100}$$

b. $3.2 \times 1.2 =$ _____

```
  3 2 tenths
×   1 2 tenths
```

c. $8.31 \times 2.4 =$ _____

d. $7.50 \times 3.5 =$ _____

4. Carolyn buys 1.2 pounds of chicken breast. If each pound of chicken breast costs $3.70, how much will she pay for the chicken breast?

5. A kitchen measures 3.75 meters by 4.2 meters.

a. Find the area of the kitchen.

b. The area of the living room is one and a half times that of the kitchen. Find the total area of the living room and the kitchen.

Lesson 18:   Relate decimal and fraction multiplication.

EUREKA MATH

Name _____ Date _____

Multiply. Do at least one problem using unit form and at least one problem using fraction form.

a. 3.2 × 1.4 =

b. 1.6 × 0.7 =

c. 2.02 × 4.2 =

d. 2.2 × 0.42 =

Angle A of a triangle is $\frac{1}{2}$ the size of angle C. Angle B is $\frac{3}{4}$ the size of angle C. If angle C measures 80 degrees, what are the measures of angle A and angle B?

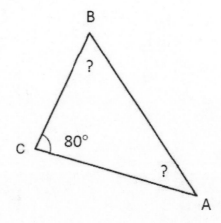

Read　　　　　**Draw**　　　　　**Write**

Lesson 19:　　Convert measures involving whole numbers, and solve multi-step word problems.

235

© 2018 Great Minds®. eureka-math.org

Name _____ Date _____

1. Convert. Express your answer as a mixed number, if possible. The first one is done for you.

---

a. 2 ft = ___$\frac{2}{3}$___ yd

$2 \text{ ft} = 2 \times 1 \text{ ft}$

$= 2 \times \frac{1}{3} \text{ yd}$

$= \frac{2}{3} \text{ yd}$

---

b. 4 ft = _____ yd

$4 \text{ ft} = 4 \times 1 \text{ ft}$

$= 4 \times$ _____ yd

$=$ _____ yd

$=$

---

c. 7 in = _____ ft

---

d. 13 in = _____ ft

---

e. 5 oz = _____ lb

---

f. 18 oz = _____ lb

Lesson 19: Convert measures involving whole numbers, and solve multi-step word problems.

© 2018 Great Minds®. eureka-math.org

2. Regina buys 24 inches of trim for a craft project.

    a. What fraction of a yard does Regina buy?

    b. If a whole yard of trim costs $6, how much did Regina pay?

3. At Yo-Yo Yogurt, the scale says that Sara has 8 ounces of vanilla yogurt in her cup. Her father's yogurt weighs 11 ounces. How many pounds of frozen yogurt did they buy altogether? Express your answer as a mixed number.

4. Pheng-Xu drinks 1 cup of milk every day for lunch. How many gallons of milk does he drink in 2 weeks?

Lesson 19:   Convert measures involving whole numbers, and solve multi-step word problems.

© 2018 Great Minds®. eureka-math.org

EUREKA
MATH

Name _____   Date _____

Convert. Express your answer as a mixed number, if possible.

a.   5 in = _____ ft

b. 13 in = _____ ft

c.   9 oz = _____ lb

d. 18 oz = _____ lb

EUREKA
MATH

Lesson 19:   Convert measures involving whole numbers, and solve multi-step
word problems.

© 2018 Great Minds®. eureka-math.org

239

A recipe calls for $\frac{3}{4}$ lb of cream cheese. A small tub of cream cheese at the grocery store weighs 12 oz. Is this enough cream cheese for the recipe?

_____

_____

_____

_____

Read   **Draw**   **Write**

Name _____   Date _____

1. Convert. Show your work. Express your answer as a mixed number. (Draw a tape diagram if it helps you.) The first one is done for you.

a. $2\frac{2}{3}$ yd = __8__ ft $2\frac{2}{3}$ yd $= 2\frac{2}{3} \times 1$ yd $\qquad = 2\frac{2}{3} \times 3$ ft $\qquad = \frac{8}{3} \times 3$ ft $\qquad = \frac{24}{3}$ ft $\qquad = 8$ ft	b. $1\frac{1}{2}$ qt = _____ gal $1\frac{1}{2}$ qt $= 1\frac{1}{2} \times 1$ qt $\qquad = 1\frac{1}{2} \times \frac{1}{4}$ gal $\qquad = \frac{3}{2} \times \frac{1}{4}$ gal $\qquad =$
c. $4\frac{2}{3}$ ft = _____ in	d. $9\frac{1}{2}$ pt = _____ qt
e. $3\frac{3}{5}$ hr = _____ min	f. $3\frac{2}{3}$ ft = _____ yd

Lesson 20:   Convert mixed unit measurements, and solve multi-step word problems.

© 2018 Great Minds®. eureka-math.org

243

2. Three dump trucks are carrying topsoil to a construction site. Truck A carries 3,545 lb, Truck B carries 1,758 lb, and Truck C carries 3,697 lb. How many tons of topsoil are the 3 trucks carrying altogether?

3. Melissa buys $3\frac{3}{4}$ gallons of iced tea. Denita buys 7 quarts more than Melissa. How much tea do they buy altogether? Express your answer in quarts.

4. Marvin buys a hose that is $27\frac{3}{4}$ feet long. He already owns a hose at home that is $\frac{2}{3}$ the length of the new hose. How many total yards of hose does Marvin have now?

Lesson 20:   Convert mixed unit measurements, and solve multi-step word problems.

© 2018 Great Minds®. eureka-math.org

EUREKA MATH

Name _____   Date _____

Convert. Express your answer as a mixed number.

a. $2\frac{1}{6}$ ft = _____ in

b. $3\frac{3}{4}$ ft = _____ yd

c. $2\frac{1}{2}$ c = _____ pt

d. $3\frac{2}{3}$ years = _____ months

EUREKA MATH®

Lesson 20:   Convert mixed unit measurements, and solve multi-step word problems.

© 2018 Great Minds®. eureka-math.org

245

Carol had $\frac{3}{4}$ yard of ribbon. She wanted to use it to decorate two picture frames. If she uses half the ribbon on each frame, how many feet of ribbon will she use for one frame? Use a tape diagram to show your thinking.

_____

_____

_____

_____

Read   **Draw**   **Write**

Lesson 21:   Explain the size of the product, and relate fraction and decimal   247
               equivalence to multiplying a fraction by 1.

© 2018 Great Minds®. eureka-math.org

Name _____   Date _____

1. Fill in the blanks. The first one has been done for you.

   a. $\frac{1}{4} \times 1 = \frac{1}{4} \times \frac{3}{3} = \frac{3}{12}$

   b. $\frac{3}{4} \times 1 = \frac{3}{4} \times \underline{} = \frac{21}{28}$

   c. $\frac{7}{4} \times 1 = \frac{7}{4} \times \underline{} = \frac{35}{20}$

   d. Use words to compare the size of the product to the size of the first factor.

2. Express each fraction as an equivalent decimal.

   a. $\frac{1}{4} \times \frac{25}{25} =$

   b. $\frac{3}{4} \times \frac{25}{25} =$

   c. $\frac{1}{5} \times \underline{} =$

   d. $\frac{4}{5} \times \underline{} =$

   e. $\frac{1}{20}$

   f. $\frac{27}{20}$

   g. $\frac{7}{4}$

   h. $\frac{8}{5}$

   i. $\frac{24}{25}$

   j. $\frac{93}{50}$

   k. $2\frac{6}{25}$

   l. $3\frac{31}{50}$

Lesson 21: Explain the size of the product, and relate fraction and decimal equivalence to multiplying a fraction by 1.

249

EUREKA MATH

© 2018 Great Minds®. eureka-math.org

3. Jack said that if you take a number and multiply it by a fraction, the product will always be smaller than what you started with. Is he correct? Why or why not? Explain your answer, and give at least two examples to support your thinking.

4. There is an infinite number of ways to represent 1 on the number line. In the space below, write at least four expressions multiplying by 1. Represent *one* differently in each expression.

5. Maria multiplied by 1 to rename $\frac{1}{4}$ as hundredths. She made factor pairs equal to 10. Use her method to change one-eighth to an equivalent decimal.

Maria's way: $\frac{1}{4} = \frac{1}{2 \times 2} \times \frac{5 \times 5}{5 \times 5} = \frac{5 \times 5}{(2 \times 5) \times (2 \times 5)} = \frac{25}{100} = 0.25$

$\frac{1}{8} =$

Paulo renamed $\frac{1}{8}$ as a decimal, too. He knows the decimal equal to $\frac{1}{4}$, and he knows that $\frac{1}{8}$ is half as much as $\frac{1}{4}$. Can you use his ideas to show another way to find the decimal equal to $\frac{1}{8}$?

Lesson 21: Explain the size of the product, and relate fraction and decimal equivalence to multiplying a fraction by 1.

EUREKA MATH

Name _____   Date _____

1. Fill in the blanks to make the equation true.

   $\frac{9}{4} \times 1 = \frac{9}{4} \times \underline{} = \frac{45}{20}$

2. Express the fractions as equivalent decimals.

   a. $\frac{1}{4} =$ b. $\frac{2}{5} =$

   c. $\frac{3}{25} =$ d. $\frac{5}{20} =$

EUREKA MATH   **Lesson 21:**   Explain the size of the product, and relate fraction and decimal   251
                                         equivalence to multiplying a fraction by 1.

                        © 2018 Great Minds®. eureka-math.org

To test her math skills, Isabella's father told her he would give her $\frac{6}{8}$ of a dollar if she could tell him how much money it is, as well as the money amount in decimal form. What should Isabella tell her father? Show your calculations.

_____

_____

_____

_____

Read   **Draw**   **Write**

Lesson 22:   Compare the size of the product to the size of the factors.

253

© 2018 Great Minds®. eureka-math.org

Name _____   Date _____

1. Solve for the unknown. Rewrite each phrase as a multiplication sentence. Circle the scaling factor and put a box around the number of meters.

   a. $\frac{1}{2}$ as long as 8 meters = _____ meter(s)

   b. 8 times as long as $\frac{1}{2}$ meter = _____ meter(s)

2. Draw a tape diagram to model each situation in Problem 1, and describe what happened to the number of meters when it was multiplied by the scaling factor.

   a.   b.

3. Fill in the blank with a numerator or denominator to make the number sentence true.

   a. $7 \times \frac{}{4} < 7$

   b. $\frac{7}{} \times 15 > 15$

   c. $3 \times \frac{}{5} = 3$

4. Look at the inequalities in each box. Choose a single fraction to write in all three blanks that would make all three number sentences true. Explain how you know.

   a.
$\frac{3}{4} \times$ ___ $> \frac{3}{4}$	$2 \times$ ___ > 2	$\frac{7}{5} \times$ ___ $> \frac{7}{5}$

   b.
$\frac{3}{4} \times$ ___ $< \frac{3}{4}$	$2 \times$ ___ < 2	$\frac{7}{5} \times$ ___ $< \frac{7}{5}$

EUREKA
MATH

Lesson 22:   Compare the size of the product to the size of the factors.

255

© 2018 Great Minds®. eureka-math.org

5. Johnny says multiplication always makes numbers bigger. Explain to Johnny why this isn't true. Give more than one example to help him understand.

6. A company uses a sketch to plan an advertisement on the side of a building. The lettering on the sketch is $\frac{3}{4}$ inch tall. In the actual advertisement, the letters must be 34 times as tall. How tall will the letters be on the building?

7. Jason is drawing the floor plan of his bedroom. He is drawing everything with dimensions that are $\frac{1}{12}$ of the actual size. His bed measures 6 ft by 3 ft, and the room measures 14 ft by 16 ft. What are the dimensions of his bed and room in his drawing?

Lesson 22:   Compare the size of the product to the size of the factors.

EUREKA MATH

Name _____ Date _____

Fill in the blank to make the number sentences true. Explain how you know.

a. $\dfrac{}{3} \times 11 > 11$

b. $5 \times \dfrac{}{8} < 5$

c. $6 \times \dfrac{2}{} = 6$

Lesson 22: Compare the size of the product to the size of the factors.

257

© 2018 Great Minds®. eureka-math.org

Jasmine took $\frac{2}{3}$ as much time to take a math test as Paula. If Paula took 2 hours to take the test, how long did it take Jasmine to take the test? Express your answer in minutes.

_____

_____

_____

_____

Read   **Draw**   **Write**

Name _____   Date _____

1. Fill in the blank using one of the following scaling factors to make each number sentence true.

1.021	0.989	1.00

   a. $3.4 \times$ _____ $= 3.4$ b. _____ $\times 0.21 > 0.21$ c. $8.04 \times$ _____ < 8.04

2.
   a. Sort the following expressions by rewriting them in the table.

The product is less than the boxed number:	The product is greater than the boxed number:

   $\boxed{13.89} \times 1.004$ $\boxed{602} \times 0.489$ $\boxed{102.03} \times 4.015$

   $\boxed{0.3} \times 0.069$ $\boxed{0.72} \times 1.24$ $\boxed{0.2} \times 0.1$

   b. Explain your sorting by writing a sentence that tells what the expressions in each column of the table have in common.

3. Write a statement using one of the following phrases to compare the value of the expressions. Then, explain how you know.

| *is slightly more than* | *is a lot more than* | *is slightly less than* | *is a lot less than* |

a. 4 × 0.988 _____ 4

b. 1.05 × 0.8 _____ 0.8

c. 1,725 × 0.013 _____ 1,725

d. 989.001 × 1.003 _____ 1.003

e. 0.002 × 0.911 _____ 0.002

Lesson 23:   Compare the size of the product to the size of the factors.

EUREKA
MATH

4. During science class, Teo, Carson, and Dhakir measure the length of their bean sprouts. Carson's sprout is 0.9 times the length of Teo's, and Dhakir's is 1.08 times the length of Teo's. Whose bean sprout is the longest? The shortest? Explain your reasoning.

5. Complete the following statements; then use decimals to give an example of each.

   ▪ $a \times b > a$ will always be true when b is...

   ▪ $a \times b < a$ will always be true when b is...

Lesson 23:   Compare the size of the product to the size of the factors.

263

© 2018 Great Minds®. eureka-math.org

Name _____ Date _____

1. Fill in the blank using one of the following scaling factors to make each number sentence true.

| 1.009 | 1.00 | 0.898 |

a. 3.06 × _____ < 3.06

b. 5.2 × _____ = 5.2

c. _____ × 0.89 > 0.89

2. Will the product of 22.65 × 0.999 be greater than or less than 22.65? Without calculating, explain how you know.

EUREKA MATH

Lesson 23: Compare the size of the product to the size of the factors.

265

© 2018 Great Minds®. eureka-math.org

Name _____   Date _____

1. A vial contains 20 mL of medicine. If each dose is $\frac{1}{8}$ of the vial, how many mL is each dose? Express your answer as a decimal.

2. A container holds 0.7 liters of oil and vinegar. $\frac{3}{4}$ of the mixture is vinegar. How many liters of vinegar are in the container? Express your answer as both a fraction and a decimal.

EUREKA
MATH®

Lesson 24:   Solve word problems using fraction and decimal multiplication.

267

© 2018 Great Minds®. eureka-math.org

3. Andres completed a 5-km race in 13.5 minutes. His sister's time was $1\frac{1}{2}$ times as long as his time. How long, in minutes, did it take his sister to run the race?

4. A clothing factory uses 1,275.2 meters of cloth a week to make shirts. How much cloth is needed to make $3\frac{3}{5}$ times as many shirts?

Lesson 24:   Solve word problems using fraction and decimal multiplication.

© 2018 Great Minds®. eureka-math.org

5. There are $\frac{3}{4}$ as many boys as girls in a class of fifth-graders. If there are 35 students in the class, how many are girls?

6. Ciro purchased a concert ticket for $56. The cost of the ticket was $\frac{4}{5}$ the cost of his dinner. The cost of his hotel was $2\frac{1}{2}$ times as much as his ticket. How much did Ciro spend altogether for the concert ticket, hotel, and dinner?

Name _____   Date _____

1. An artist builds a sculpture out of metal and wood that weighs 14.9 kilograms. $\frac{3}{4}$ of this weight is metal, and the rest is wood. How much does the wood part of the sculpture weigh?

2. On a boat tour, there are half as many children as there are adults. There are 30 people on the tour. How many children are there?

Lesson 24:   Solve word problems using fraction and decimal multiplication.

© 2018 Great Minds®. eureka-math.org

271

The label on a 0.118 L bottle of cough syrup recommends a dose of 10 mL for children aged 6 to 10 years. How many 10 mL doses are in the bottle?

_____

_____

_____

_____

Read   **Draw**   **Write**

Name _____   Date _____

1. Draw a tape diagram and a number line to solve. You may draw the model that makes the most sense to you. Fill in the blanks that follow. Use the example to help you.

   Example: $2 \div \frac{1}{3} = $ __6__

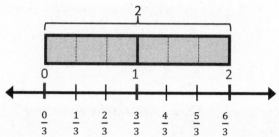   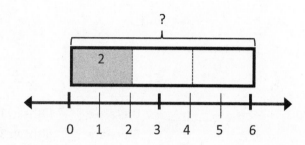

   There are __3__ thirds in 1 whole.   If 2 is $\frac{1}{3}$, what is the whole? ___6___

   There are __6__ thirds in 2 wholes.

   a. $4 \div \frac{1}{2} = $ _____   There are ____ halves in 1 whole.   If 4 is $\frac{1}{2}$, what is the whole? _____
                                            There are ____ halves in 4 wholes.

   b. $2 \div \frac{1}{4} = $ _____   There are ____ fourths in 1 whole.   If 2 is $\frac{1}{4}$, what is the whole? _____
                                            There are ____ fourths in 2 wholes.

c. $5 \div \frac{1}{3} =$ _____   There are ____ thirds in 1 whole.   If 5 is $\frac{1}{3}$, what is the whole? _____

There are ____ thirds in 5 wholes.

d. $3 \div \frac{1}{3} =$ _____   There are ____ fifths in 1 whole.   If 3 is $\frac{1}{5}$, what is the whole? _____

There are ____ fifths in 3 wholes.

2. Divide. Then, multiply to check.

a. $5 \div \frac{1}{2}$	b. $3 \div \frac{1}{2}$	c. $4 \div \frac{1}{5}$	d. $1 \div \frac{1}{6}$
e. $2 \div \frac{1}{8}$	f. $7 \div \frac{1}{6}$	g. $8 \div \frac{1}{3}$	h. $9 \div \frac{1}{4}$

Lesson 25:   Divide a whole number by a unit fraction.

EUREKA MATH

3. For an art project, Mrs. Williams is dividing construction paper into fourths. How many fourths can she make from 5 pieces of construction paper?

4. Use the chart below to answer the following questions.

Donnie's Diner Lunch Menu

Food	Serving Size
Hamburger	$\frac{1}{3}$ lb
Pickles	$\frac{1}{4}$ pickle
Potato chips	$\frac{1}{8}$ bag
Chocolate milk	$\frac{1}{2}$ cup

a. How many hamburgers can Donnie make with 6 pounds of hamburger meat?

b. How many pickle servings can be made from a jar of 15 pickles?

c.   How many servings of chocolate milk can he serve from a gallon of milk?

5.   Three gallons of water fill $\frac{1}{4}$ of the elephant's pail at the zoo. How much water does the pail hold?

Lesson 25:   Divide a whole number by a unit fraction.

EUREKA
MATH

Name _____   Date _____

1. Draw a tape diagram and a number line to solve. Fill in the blanks that follow.

    a. $5 \div \frac{1}{2} =$ _____

                                      There are ____ halves in 1 whole.

                                      There are ____ halves in 5 wholes.

                                      5 is $\frac{1}{2}$ of what number? _____

    b. $4 \div \frac{1}{4} =$ _____

                                        There are ____ fourths in 1 whole.

                                      There are ____ fourths in ____ wholes.

                                      4 is $\frac{1}{4}$ of what number? _____

2. Ms. Leverenz is doing an art project with her class. She has a 3 foot piece of ribbon. If she gives each student an eighth of a foot of ribbon, will she have enough for her class of 22 students?

A race begins with $2\frac{1}{2}$ miles through town, continues through the park for $2\frac{1}{3}$ miles, and finishes at the track after the last $\frac{1}{6}$ mile. A volunteer is stationed every quarter mile and at the finish line to pass out cups of water and cheer on the runners. How many volunteers are needed?

_____

_____

_____

_____

Read　　　　　　**Draw**　　　　　　**Write**

EUREKA MATH

Name _____   Date _____

1. Draw a model or tape diagram to solve. Use the thought bubble to show your thinking. Write your quotient in the blank. Use the example to help you.

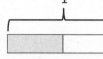

Example: $\frac{1}{2} \div 3$

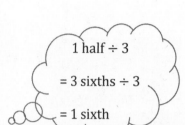

1 half ÷ 3

= 3 sixths ÷ 3

= 1 sixth

$\frac{1}{2} \div 3 = \frac{1}{6}$

a. $\frac{1}{3} \div 2 =$ _____

b. $\frac{1}{3} \div 4 =$ _____

EUREKA
MATH®

c. $\frac{1}{4} \div 2 =$ _____

d. $\frac{1}{4} \div 3 =$ _____

2. Divide. Then, multiply to check.

a. $\frac{1}{2} \div 7$	b. $\frac{1}{3} \div 6$	c. $\frac{1}{4} \div 5$	d. $\frac{1}{5} \div 4$
e. $\frac{1}{5} \div 2$	f. $\frac{1}{6} \div 3$	g. $\frac{1}{8} \div 2$	h. $\frac{1}{10} \div 10$

Lesson 26: Divide a unit fraction by a whole number.

EUREKA MATH

3. Tasha eats half her snack and gives the other half to her two best friends for them to share equally. What portion of the whole snack does each friend get? Draw a picture to support your response.

4. Mrs. Appler used $\frac{1}{2}$ gallon of olive oil to make 8 identical batches of salad dressing.

   a. How many gallons of olive oil did she use in each batch of salad dressing?

   b. How many cups of olive oil did she use in each batch of salad dressing?

5. Mariano delivers newspapers. He always puts $\frac{3}{4}$ of his weekly earnings in his savings account and then divides the rest equally into 3 piggy banks for spending at the snack shop, the arcade, and the subway.

  a. What fraction of his earnings does Mariano put into each piggy bank?

  b. If Mariano adds $2.40 to each piggy bank every week, how much does Mariano earn per week delivering papers?

Lesson 26:   Divide a unit fraction by a whole number.

EUREKA MATH

Name _____   Date _____

1. Solve. Support at least one of your answers with a model or tape diagram.

   a. $\frac{1}{2} \div 4 =$ _____

   b. $\frac{1}{8} \div 5 =$ _____

2. Larry spends half of his workday teaching piano lessons. If he sees 6 students, each for the same amount of time, what fraction of his workday is spent with each student?

Name _____   Date _____

1. Mrs. Silverstein bought 3 mini cakes for a birthday party. She cuts each cake into quarters and plans to serve each guest 1 quarter of a cake. How many guests can she serve with all her cakes? Draw a picture to support your response.

2. Mr. Pham has $\frac{1}{4}$ pan of lasagna left in the refrigerator. He wants to cut the lasagna into equal slices so he can have it for dinner for 3 nights. How much lasagna will he eat each night? Draw a picture to support your response.

3. The perimeter of a square is $\frac{1}{5}$ of a meter.

   a. Find the length of each side in meters. Draw a picture to support your response.

   b. How long is each side in centimeters?

4. A pallet holding 5 identical crates weighs $\frac{1}{4}$ of a ton.

   a. How many tons does each crate weigh? Draw a picture to support your response.

EUREKA MATH

b. How many pounds does each crate weigh?

5. Faye has 5 pieces of ribbon, each 1 yard long. She cuts each ribbon into sixths.

a. How many sixths will she have after cutting all the ribbons?

b. How long will each of the sixths be in inches?

6. A glass pitcher is filled with water. $\frac{1}{8}$ of the water is poured equally into 2 glasses.

    a. What fraction of the water is in each glass?

    b. If each glass has 3 fluid ounces of water in it, how many fluid ounces of water were in the full pitcher?

    c. If $\frac{1}{4}$ of the remaining water is poured out of the pitcher to water a plant, how many cups of water are left in the pitcher?

Lesson 27:   Solve problems involving fraction division.

EUREKA
MATH

Name _____ Date _____

1. Kevin divides 3 pieces of paper into fourths. How many fourths does he have? Draw a picture to support your response.

2. Sybil has $\frac{1}{2}$ of a pizza left over. She wants to share the pizza with 3 of her friends. What fraction of the original pizza will Sybil and her 3 friends each receive? Draw a picture to support your response.

Name _____   Date _____

1. Create and solve a division story problem about 5 meters of rope that is modeled by the tape diagram below.

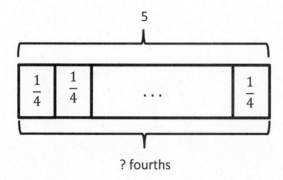

? fourths

2. Create and solve a story problem about $\frac{1}{4}$ pound of almonds that is modeled by the tape diagram below.

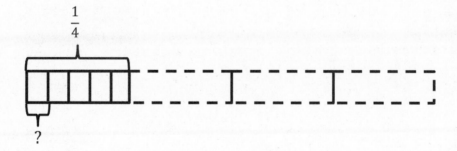

Lesson 28:   Write equations and word problems corresponding to tape and number line diagrams.

295

3. Draw a tape diagram and create a word problem for the following expressions, and then solve.

a. $2 \div \frac{1}{3}$

b. $\frac{1}{3} \div 4$

c. $\frac{1}{4} \div 3$

d. $3 \div \frac{1}{5}$

Lesson 28:   Write equations and word problems corresponding to tape and
number line diagrams.

EUREKA
MATH

Name _____ Date _____

Create a word problem for the following expressions, and then solve.

a. $4 \div \frac{1}{2}$

b. $\frac{1}{2} \div 4$

Lesson 28:   Write equations and word problems corresponding to tape and number line diagrams.

© 2018 Great Minds®. eureka-math.org

297

Fernando bought a jacket for \$185 and sold it for $1\frac{1}{2}$ times what he paid. Marisol spent $\frac{1}{5}$ as much as Fernando on the same jacket but sold it for $\frac{1}{2}$ as much as Fernando did. How much money did Marisol make? Explain your thinking using a diagram.

_____

_____

_____

_____

Read   **Draw**   **Write**

EUREKA MATH

Lesson 29:   Connect division by a unit fraction to division by 1 tenth and 1 hundredth.

299

Name _____   Date _____

1. Divide. Rewrite each expression as a division sentence with a fraction divisor, and fill in the blanks. The first one is done for you.

Example: $2 \div 0.1 = 2 \div \frac{1}{10} = 20$

There are ___10___ tenths in 1 whole.

There are ___20___ tenths in 2 wholes.

a. $5 \div 0.1$

There are _____ tenths in 1 whole.

There are _____ tenths in 5 wholes.

b. $8 \div 0.1$

There are _____ tenths in 1 whole.

There are _____ tenths in 8 wholes.

c. $5.2 \div 0.1$

There are _____ tenths in 5 wholes.

There are _____ tenths in 2 tenths.

There are _____ tenths in 5.2.

d. $8.7 \div 0.1$

There are _____ tenths in 8 wholes.

There are _____ tenths in 7 tenths.

There are _____ tenths in 8.7.

e. $5 \div 0.01$

There are _____ hundredths in 1 whole.

There are _____ hundredths in 5 wholes.

f. $8 \div 0.01$

There are _____ hundredths in 1 whole.

There are _____ hundredths in 8 wholes.

g. $5.2 \div 0.01$

There are _____ hundredths in 5 wholes.

There are _____ hundredths in 2 tenths.

There are _____ hundredths in 5.2.

h. $8.7 \div 0.01$

There are _____ hundredths in 8 wholes.

There are _____ hundredths in 7 tenths.

There are _____ hundredths in 8.7.

2. Divide.

a. 6 ÷ 0.1	b. 18 ÷ 0.1	c. 6 ÷ 0.01
d. 1.7 ÷ 0.1	e. 31 ÷ 0.01	f. 11 ÷ 0.01
g. 125 ÷ 0.1	h. 3.74 ÷ 0.01	i. 12.5 ÷ 0.01

3. Yung bought $4.60 worth of bubble gum. Each piece of gum cost $0.10. How many pieces of bubble gum did Yung buy?

4. Cheryl solved a problem: 84 ÷ 0.01 = 8,400.

   Jane said, "Your answer is wrong because when you divide, the quotient is always smaller than the whole amount you start with, for example, 6 ÷ 2 = 3 and 100 ÷ 4 = 25." Who is correct? Explain your thinking.

5. The U.S. Mint sells 2 ounces of American Eagle gold coins to a collector. Each coin weighs one-tenth of an ounce. How many gold coins were sold to the collector?

Lesson 29:   Connect division by a unit fraction to division by 1 tenth and
1 hundredth.

© 2018 Great Minds®. eureka-math.org

EUREKA
MATH

Name _____   Date _____

1. 8.3 is equal to

   _____ tenths

   _____ hundredths

2. 28 is equal to

   _____ hundredths

   _____ tenths

3. 15.09 ÷ 0.01 = _____

4. $267.4 ÷ \frac{1}{10}$ = _____

5. $632.98 ÷ \frac{1}{100}$ = _____

EUREKA MATH

Lesson 29:   Connect division by a unit fraction to division by 1 tenth and
1 hundredth.

© 2018 Great Minds®. eureka-math.org

303

Alexa claims that $16 \div 4$, $\frac{32}{8}$, and 8 halves are all equivalent expressions. Is Alexa correct? Explain how you know.

_____

_____

_____

_____

Read   **Draw**   **Write**

Name _____   Date _____

1. Rewrite the division expression as a fraction and divide. The first two have been started for you.

a. $2.7 \div 0.3 = \dfrac{2.7}{0.3}$ $= \dfrac{2.7 \times 10}{0.3 \times 10}$ $= \dfrac{27}{3}$ $= 9$	b. $2.7 \div 0.03 = \dfrac{2.7}{0.03}$ $= \dfrac{2.7 \times 100}{0.03 \times 100}$ $= \dfrac{270}{3}$ $=$
c. $3.5 \div 0.5$	d. $3.5 \div 0.05$
e. $4.2 \div 0.7$	f. $0.42 \div 0.07$

g. $10.8 \div 0.9$	h. $1.08 \div 0.09$
i. $3.6 \div 1.2$	j. $0.36 \div 0.12$
k. $17.5 \div 2.5$	l. $1.75 \div 0.25$

2. $15 \div 3 = 5$. Explain why it is true that $1.5 \div 0.3$ and $0.15 \div 0.03$ have the same quotient.

Lesson 30:   Divide decimal dividends by non-unit decimal divisors.

© 2018 Great Minds®. eureka-math.org

EUREKA
MATH

3. Mr. Volok buys 2.4 kg of sugar for his bakery.

    a. If he pours 0.2 kg of sugar into separate bags, how many bags of sugar can he make?

    b. If he pours 0.4 kg of sugar into separate bags, how many bags of sugar can he make?

4. Two wires, one 17.4 meters long and one 7.5 meters long, were cut into pieces 0.3 meters long. How many such pieces can be made from both wires?

5. Mr. Smith has 15.6 pounds of oranges to pack for shipment. He can ship 2.4 pounds of oranges in a large box and 1.2 pounds in a small box. If he ships 5 large boxes, what is the minimum number of small boxes required to ship the rest of the oranges?

Lesson 30:   Divide decimal dividends by non-unit decimal divisors.

309

© 2018 Great Minds®. eureka-math.org

Name _____   Date _____

Rewrite the division expression as a fraction and divide.

a. $3.2 \div 0.8$	b. $3.2 \div 0.08$
c. $7.2 \div 0.9$	d. $0.72 \div 0.09$

A café makes ten 8-ounce fruit smoothies. Each smoothie is made with 4 ounces of soy milk and 1.3 ounces of banana flavoring. The rest is blueberry juice. How much of each ingredient will be necessary to make the smoothies?

Read   **Draw**   **Write**

EUREKA MATH

Lesson 31:   Divide decimal dividends by non-unit decimal divisors.

313

© 2018 Great Minds®. eureka-math.org

Name _____   Date _____

1. Estimate and then divide. An example has been done for you.

$78.4 \div 0.7 \approx 770 \div 7 = 110$

$= \dfrac{78.4}{0.7}$

$= \dfrac{78.4 \times 10}{0.7 \times 10}$

$= \dfrac{784}{7}$

$= 112$

```
       1 1 2
   7 | 7 8 4
      -7
       8
      -7
       1 4
      -1 4
         0
```

     a. $53.2 \div 0.4 \approx$ b. $1.52 \div 0.8 \approx$

2. Estimate and then divide. The first one has been done for you.

$7.32 \div 0.06 \approx 720 \div 6 = 120$

$= \dfrac{7.32}{0.06}$

$= \dfrac{7.32 \times 100}{0.06 \times 100}$

$= \dfrac{732}{6}$

$= 122$

```
       1 2 2
   6 | 7 3 2
      -6
       1 3
      -1 2
       1 2
      -1 2
         0
```

     a. $9.42 \div 0.03 \approx$ b. $39.36 \div 0.96 \approx$

3. Solve using the standard algorithm. Use the thought bubble to show your thinking as you rename the divisor as a whole number.

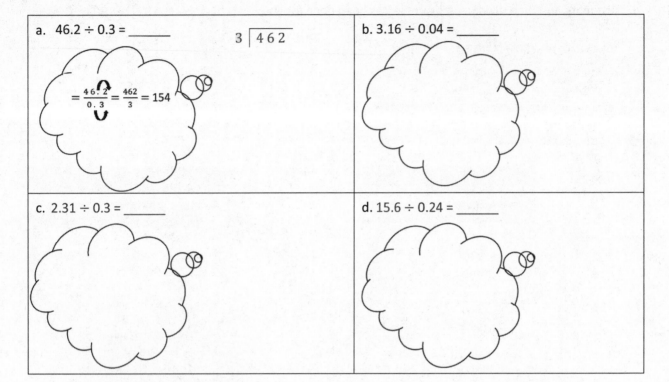

a. $46.2 \div 0.3 = $ _____

$$3 \overline{\smash{)}462}$$

$$= \frac{46.2}{0.3} = \frac{462}{3} = 154$$

b. $3.16 \div 0.04 = $ _____

c. $2.31 \div 0.3 = $ _____

d. $15.6 \div 0.24 = $ _____

4. The total distance of a race is 18.9 km.

   a. If volunteers set up a water station every 0.7 km, including one at the finish line, how many stations will they have?

   b. If volunteers set up a first aid station every 0.9 km, including one at the finish line, how many stations will they have?

5. In a laboratory, a technician combines a salt solution contained in 27 test tubes. Each test tube contains 0.06 liter of the solution. If he divides the total amount into test tubes that hold 0.3 liter each, how many test tubes will he need?

        Lesson 31:   Divide decimal dividends by non-unit decimal divisors.   **EUREKA MATH**

© 2018 Great Minds®. eureka-math.org

Name _____   Date _____

Estimate first, and then solve using the standard algorithm. Show how you rename the divisor as a whole number.

1. 6.39 ÷ 0.09

2. 82.14 ÷ 0.6

Lesson 31:   Divide decimal dividends by non-unit decimal divisors.

317

© 2018 Great Minds®. eureka-math.org

Four baby socks can be made from $\frac{1}{3}$ skein of yarn. How many baby socks can be made from a whole skein? Draw a model to show your thinking.

_____

_____

_____

_____

Read   **Draw**   **Write**

Lesson 32:   Interpret and evaluate numerical expressions including the language of scaling and fraction division.

319

© 2018 Great Minds®. eureka-math.org

Name _____   Date _____

1. Circle the expression equivalent to *the sum of 3 and 2 divided by* $\frac{1}{3}$.

   $\dfrac{3+2}{3}$ $3 + (2 \div \frac{1}{3})$ $(3 + 2) \div \frac{1}{3}$ $\frac{1}{3} \div (3 + 2)$

2. Circle the expression(s) equivalent to *28 divided by the difference between* $\frac{4}{5}$ *and* $\frac{7}{10}$.

   $28 \div \left(\frac{4}{5} - \frac{7}{10} \right)$ $\dfrac{28}{\frac{4}{5} - \frac{7}{10}}$ $\left(\frac{4}{5} - \frac{7}{10} \right) \div 28$ $28 \div \left(\frac{7}{10} - \frac{4}{5} \right)$

3. Fill in the chart by writing an equivalent numerical expression.

a.	Half as much as the difference between $2\frac{1}{4}$ and $\frac{3}{8}$.	
b.	The difference between $2\frac{1}{4}$ and $\frac{3}{8}$ divided by 4.	
c.	A third of the sum of $\frac{7}{8}$ and 22 tenths.	
d.	Add 2.2 and $\frac{7}{8}$, and then triple the sum.	

4. Compare expressions 3(a) and 3(b). Without evaluating, identify the expression that is greater. Explain how you know.

EUREKA
MATH

Lesson 32:   Interpret and evaluate numerical expressions including the language of scaling and fraction division.

321

© 2018 Great Minds®. eureka-math.org

5. Fill in the chart by writing an equivalent expression in word form.

a.		$\frac{3}{4} \times (1.75 + \frac{3}{5})$
b.		$\frac{7}{9} - (\frac{1}{8} \times 0.2)$
c.		$(1.75 + \frac{3}{5}) \times \frac{4}{3}$
d.		$2 \div (\frac{1}{2} \times \frac{4}{5})$

6. Compare the expressions in 5(a) and 5(c). Without evaluating, identify the expression that is less. Explain how you know.

7. Evaluate the following expressions.

a. $(9 - 5) \div \frac{1}{3}$

b. $\frac{5}{3} \times (2 \times \frac{1}{4})$

c. $\frac{1}{3} \div (1 \div \frac{1}{4})$

Lesson 32:   Interpret and evaluate numerical expressions including the language of scaling and fraction division.

EUREKA MATH

d. $\frac{1}{2} \times \frac{3}{5} \times \frac{5}{3}$ e. Half as much as $(\frac{3}{4} \times 0.2)$ f. 3 times as much as the quotient of 2.4 and 0.6

8. Choose an expression below that matches the story problem, and write it in the blank.

$\frac{2}{3} \times (20 - 5)$ $(\frac{2}{3} \times 20) - (\frac{2}{3} \times 5)$ $\frac{2}{3} \times 20 - 5$ $(20 - \frac{2}{3}) - 5$

a. Farmer Green picked 20 carrots. He cooked $\frac{2}{3}$ of them, and then gave 5 to his rabbits. Write the expression that tells how many carrots he had left.

Expression: _____

b. Farmer Green picked 20 carrots. He cooked 5 of them, and then gave $\frac{2}{3}$ of the remaining carrots to his rabbits. Write the expression that tells how many carrots the rabbits will get.

Expression: _____

Name _____ Date _____

1. Write an equivalent expression in numerical form.

   A fourth as much as the product of two-thirds and 0.8

2. Write an equivalent expression in word form.

   a. $\frac{3}{8} \times (1 - \frac{1}{3})$

   b. $(1 - \frac{1}{3}) \div 2$

3. Compare the expressions in 2(a) and 2(b). Without evaluating, determine which expression is greater, and explain how you know.

EUREKA MATH

Lesson 32: Interpret and evaluate numerical expressions including the language of scaling and fraction division.

© 2018 Great Minds®. eureka-math.org

325

Name _____ Date _____

1. Ms. Hayes has $\frac{1}{2}$ liter of juice. She distributes it equally to 6 students in her tutoring group.

    a. How many liters of juice does each student get?

    b. How many more liters of juice will Ms. Hayes need if she wants to give each of the 24 students in her class the same amount of juice found in Part (a)?

2. Lucia has 3.5 hours left in her workday as a car mechanic. Lucia needs $\frac{1}{2}$ of an hour to complete one oil change.

    a. How many oil changes can Lucia complete during the rest of her workday?

    b. Lucia can complete two car inspections in the same amount of time it takes her to complete one oil change. How long does it take her to complete one car inspection?

    c. How many inspections can she complete in the rest of her workday?

Lesson 33: Create story contexts for numerical expressions and tape diagrams, and solve word problems.

327

3. Carlo buys $14.40 worth of grapefruit. Each grapefruit costs $0.80.

   a. How many grapefruits does Carlo buy?

   b. At the same store, Kahri spends one-third as much money on grapefruits as Carlo. How many grapefruits does she buy?

4. Studies show that a typical giant hummingbird can flap its wings once in 0.08 of a second.

   a. While flying for 7.2 seconds, how many times will a typical giant hummingbird flap its wings?

   b. A ruby-throated hummingbird can flap its wings 4 times faster than a giant hummingbird. How many times will a ruby-throated hummingbird flap its wings in the same amount of time?

Lesson 33:   Create story contexts for numerical expressions and tape diagrams, and solve word problems.

EUREKA
MATH

5. Create a story context for the following expression.

$$\frac{1}{3} \times (\$20 - \$3.20)$$

6. Create a story context about painting a wall for the following tape diagram.

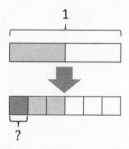

Lesson 33:   Create story contexts for numerical expressions and tape diagrams,
and solve word problems.

© 2018 Great Minds®. eureka-math.org

Name _____ Date _____

An entire commercial break is 3.6 minutes.

    a. If each commercial takes 0.6 minutes, how many commercials will be played?

    b. A different commercial break of the same length plays commercials half as long. How many commercials will play during this break?

Lesson 33:   Create story contexts for numerical expressions and tape diagrams, and solve word problems.

331

© 2018 Great Minds®. eureka-math.org

Credits

Great Minds® has made every effort to obtain permission for the reprinting of all copyrighted material. If any owner of copyrighted material is not acknowledged herein, please contact Great Minds for proper acknowledgment in all future editions and reprints of this module.

© 2018 Great Minds®. eureka-math.org